Lecture Notes in Bioengineering

Lecture Notes in Bioengineering (LNBE) publishes the latest developments in bioengineering. It covers a wide range of topics, including (but not limited to):

- Bio-inspired Technology & Biomimetics
- Biosensors
- Bionanomaterials
- Biomedical Instrumentation
- Biological Signal Processing
- Medical Robotics and Assistive Technology
- Computational Medicine, Computational Pharmacology and Computational Biology
- Personalized Medicine
- Data Analysis in Bioengineering
- Neuroengineering
- Bioengineering Ethics

Original research reported in proceedings and edited books are at the core of LNBE. Monographs presenting cutting-edge findings, new perspectives on classical fields or reviewing the state-of-the art in a certain subfield of bioengineering may exceptionally be considered for publication. Alternatively, they may be redirected to more specific book series. The series' target audience includes advanced level students, researchers, and industry professionals working at the forefront of their fields.

Indexed by SCOPUS and Springerlink. The books of the series are submitted for indexing to Web of Science.

More information about this series at http://www.springer.com/series/11564

Samira Hosseini ·
Michelle Alejandra Espinosa-Hernandez ·
Ricardo Garcia-Ramirez ·
Ana Sofia Cerda-Kipper ·
Sofia Reveles-Huizar ·
Luis Acosta-Soto

BioMEMS

Biosensing Applications

Springer

Samira Hosseini
School of Engineering and Sciences
Tecnologico de Monterrey
Monterrey, Mexico

Michelle Alejandra Espinosa-Hernandez
School of Engineering and Sciences
Tecnologico de Monterrey
Monterrey, Mexico

Ricardo Garcia-Ramirez
School of Engineering and Sciences
Tecnologico de Monterrey
Monterrey, Mexico

Ana Sofia Cerda-Kipper
School of Engineering and Sciences
Tecnologico de Monterrey
Monterrey, Mexico

Sofia Reveles-Huizar
School of Engineering and Sciences
Tecnologico de Monterrey
Monterrey, Mexico

Luis Acosta-Soto
School of Engineering and Sciences
Tecnologico de Monterrey
Monterrey, Mexico

ISSN 2195-271X ISSN 2195-2728 (electronic)
Lecture Notes in Bioengineering
ISBN 978-981-15-6381-2 ISBN 978-981-15-6382-9 (eBook)
https://doi.org/10.1007/978-981-15-6382-9

This Springer imprint is published by the registered company Springer Nature Singapore Pte Ltd.
The registered company address is: 152 Beach Road, #21-01/04 Gateway East, Singapore 189721, Singapore

Preface

Over the past few decades, biological microelectro-mechanical systems (BioMEMS) have been commonly designed, fabricated, and used for a wide range of applications including cell studies, therapeutics, tissue engineering, drug delivery, implantable devices, and biosensors, among others. BioMEMS are portable, low-cost, rapid, and robust platforms designed to automate one or more analysis steps such as mixing, separation, sedimentation, etc. into one monolithic device. BioMEMS have proven to be excellent candidates for biosensing applications. Such devices act as miniaturized laboratory systems that can be used in remote and/or rural areas for detection of multiple analytes with minimal human involvement, which, in turn, makes the detection procedure of fatal diseases safer for laboratory technicians. This book is dedicated to the latest advancements of BioMEMS in biosensing applications. Different detection strategies including colorimetric, fluorescence, luminescence, bioluminescence, chemiluminescence, biochemiluminescence, and electrochemiluminescence are thoroughly reviewed in this book, and different types of BioMEMS designed and fabricated for the mentioned detection strategies are presented with recent examples. These BioMEMS devices include paper-based, microfluidics such as lab-on-chip (LOC), lab-on-compact-disk (LOCD) systems and interesting alternative techniques that offer new solutions. This book also provides an overview on the history of BioMEMS and the pioneered devices in the history of science which were designed and fabricated for different purposes.

Monterrey, Mexico

Samira Hosseini
Ricardo Garcia-Ramirez
Michelle Alejandra Espinosa-Hernandez
Sofia Reveles-Huizar
Luis Acosta-Soto
Ana Sofia Cerda-Kipper

Acknowledgements The authors would like to acknowledge the financial and technical support of Writing Lab, TecLabs, Tecnologico de Monterrey, in the production of this work. The authors would also like to acknowledge the kind support of Dr. Aida Rodriquez-Garcia and Mrs. Niousha Mousavi in editing this work.

Contents

Abbreviations

2D	Two dimensional
3D	Three dimensional
ABTS	2,2′-Azino-bis (3-ethylbenz-thiazoline-6-sulfonic acid)
ACh	Acetylcholine
AChE	Acetylcholinesterase
acpcPNA	Pyrrolidinyl peptide nucleic acid
ADCs	Analog-to-digital converters
AgNPs	Silver nanoparticles
Al	Aluminum
ALP	Alkaline phosphatase
AMPPD	3-(2′-Spiroadamantyl)-4-methoxy-4-(3″-phosphoryloxy) phenyl-1,2-dioxetane
Ann-V	Annexin V
anti-hCGα	Anti-human chorionic gonadotropin Gα
AP	Alkaline phosphatase
APTES	(3-Aminopropyl) tri-ethoxysilane
a-Si:H	Hydrogenated amorphous silicon
a-SiN_x	Amorphous silicon nitride
AST	Antimicrobial susceptibility testing
ATChI	Acetylthiocholine iodide
ATP	Adenosine 5′-triphosphate
ATP-BLA	ATP bioluminescence assay
AuNC	Gold nanoclusters
AuNPs	Gold nanoparticles
BioMEMS	Biological microelectro-mechanical systems
BL	Bioluminescence
BL-CL	Biochemiluminescence
BSA	Bovine serum albumin
BuChE	Butyrylcholinesterase
BuTChI	Butyrylthiocholine iodide

CAMPT	Camptothecin
CBM	Carbohydrate binding module
CCD	Charge-coupled device
CCGTSs	Chemiluminescence cloth-based glucose test sensors
CD	Compact disk
C-dots	Carbon dots
C-dots@NPG	Carbon dots dotted nanoporous gold
CL	Chemiluminescence
CMG	Water-soluble carboxymethylated β-1,3-glucan
CMOS	Complementary metal oxide semiconductor
CMVs	"Culture medium veins"
CNC	Computer numerical control
CRP	C-reactive protein
Cu	Copper
CVD	Chemical vapor deposition
C-μPAD	Chemically patterned μPAD
DC	Direct current
DDS	Drug delivery system
DED	Dry eye disease
DHBS	2-Hydroxy-3,5-dichlorobenzenesulfonic acid
DNA	Deoxyribonucleic acid
DNase	Deoxyribonuclease
DO	Dissolved oxygen
DRIE	Deep reaction ion etching
DSC	N,N'-disuccinimidyl carbonate
E. Coli	Escherichia coli
ECL	Electrochemiluminescence
EDC	1-Ethyl-3-(3-dimethylaminopropyl) carbodiimide
e-ELISA	Electronics-based enzyme-linked immunosorbent assay
ELISA	Enzyme-linked immunosorbent assay
EMVs	Electromagnetic valves
EPOC	Extreme point of care
Exo	Exonuclease
FEP	Fluorinated ethylene propylene
FITC	Fluorescein isothiocyanate
FMN	Flavin mononucleotide
FRET	Förster resonance energy transfer
FTA	Fast technology analysis
GDNA	Guanine-rich DNA
GLU	Glucose
GOx	Glucose oxidase
GQD	Graphene quantum dots
GS	Gas chromatography
H_2O_2	Hydrogen peroxide
HbA1c	Blood glycated hemoglobin

HBsAg	Hepatitis B surface antigen
hCG	Human chorionic gonadotropin
HIV	Human immunodeficiency virus
HPV	Human papillomavirus
HRP	Horseradish peroxidase
HSL	Hue–saturation–lightness
IATP-BLA	Immunosorbent ATP-bioluminescence assay
icELISA	Indirect competitive enzyme-linked immunosorbent assay
IFNγ	Interferon gamma
IgG	Immunoglobulin G
IgY	Immunoglobulin Y
ISS	International Space Station
ITO	Indium tin oxide
KPC	*Klebsiella pneumoniae carbapenemase*
LAMP	Loop-mediated isothermal amplification
LED	Light-emitting diode
LFIA	Lateral flow immunoassay
LOC	Lab-on-chip
LOD	Limit of detection
LoPCB	Lab-on-a-printed circuit board
LPCVD	Low-pressure chemical vapor deposition
LSA	Long spacer arm
LSA-MPs	Long spacer arm-functionalized magnetic particles
MB	Molecular beacon
MDMHA	3,4-Ethylenedioxy-N-methylamphetamine
MEMS/MOEMS	Micro-(opto)-electromechanical systems
MERS	Middle East respiratory syndrome
MERS-CoV	Middle East respiratory syndrome coronavirus
miRNA	MicroRNA
MNC-EC	MNC-E. Coli
MNCs	Magnetic nanoparticle clusters
MNPs	Magnetic nanoparticles
MORP	Morpholinopyridine
MOTiF	Multi-organ-tissue-flow
MPA	3-Mercaptopropionic acid
MPs	Magnetic particles
mRNA	Messenger ribonucleic acid
NC	Nanocluster
NHS	N-hydroxysuccinimide
NP	Nanoparticle(s)
NPG	Nanoporous gold
O_2	Oxygen
OFSE	Oral fluid sampling equipment
OLED	Organic electroluminescent diode

OPD	Organic photodiode
OPs	Organophosphate pesticides
OTA	Ochratoxin A
PAD	Paper-based analytical devices
PBS	Phosphate buffer saline
PCB	Printed circuit boards
PCR	Polymerase chain reaction
PDMS	Polydimethylsiloxane
PECVD	Plasma-enhanced chemical vapor deposition
PEDOT	Poly(3,4-ethylenedioxythiopene)
PEG	Poly(ethylene glycol)
PEGDA	Polyethylene-glycol diacrylate
P-ELISAs	Paper-based enzyme-linked immunosorbent assays
PGMEA	Propylene glycol methyl ether acetate
PIF	Parity inner fails
PIP	P-iodophenol
PLSR	Partial least square regression
PMMA	Poly(methyl methacrylate)
POC	Point of care
POCT	Point-of-care testing
Poly(HEMA-co-AEMA)	Poly(2-hydroxyethyl methacrylate-co-2-aminoethyl methacrylate)
PP	Polypropylene
Ps	Phosphatidylserine
PS	Polystyrene
PSA	Pressure-sensitive adhesive
Pt	Platinum
PtOEP	2,3,7,8,12,13,17,18-Octaethyl-21H,23H-porphyrin, platinum (II)
PVC	Hydrophobic polyvinyl chloride
RCA	Rolling circle amplification
RhB	Rhodamine B
RhB-ITC	Rhodamine B isothiocyanate
RIE	Reactive ion etching
RNA	Ribonucleic acid
RPM	Revulsion per minute
RT-PCR	Reverse transcription PCR
SA-AP	Streptavidin–alkaline phosphatase
SDA	Strand displacement amplification
SiPMs	Silicon photomultipliers
SLA	Stereolithography
SmartBA	Total bile acid smartphone-based assay
SmartChol	Total cholesterol smartphone-based assay
SPTZ	3-(10'-Phenothiazinyl)propane-1-sulfonate
SSC	Saline sodium citrate

ssDNA	Single-stranded DNA
T1D	Type 1 diabetes
Tb	Tuberculosis
TC	Tetracycline
TC	Total cholesterol
$tcpO_2$	Transcutaneous O_2 pressure
TCS	Trichlorosilane
TEOS	Tetraethyl orthosilicate
TG	Triglyceride
TiO_2 NPs	Titanium dioxide nanoparticles
TIRCA	Toehold-initiated RCA
TiW	Titanium tungsten
TLC	Thin-layer chromatography
TMB	3,3',5,5'-Tetramethylbenzidine
TNFα	Tumor necrosis factor alpha
ULOC	Unibody-LOC
UTI	Urinary tract infection
UV	Ultraviolet
UV–vis	Ultraviolet–visible
WFP	Whatman filter paper
ZnO-NRs	Zinc oxide nanorods
μCAD	Cloth-based analytical devices
μPADs	Microfluidic paper analytical devices
μ-TAS	Micro-total analysis systems

Chapter 1
History of Bio-microelectromechanical Systems (BioMEMS)

Ricardo Garcia-Ramirez and Samira Hosseini

1.1 Introduction

MEMS are miniaturized devices that transduce signals in different domains using semiconductor manufacturing techniques to produce non-electrical elements (Council 1998). The acronym MEMS was coined in the United States before the start of the 1990s by Professor Roger Thomas Howe, along with other scientists, after microscale fabrication was established as a growing engineering field (Maluf and Williams 2004). The first conference regarding MEMS was held in Berlin in 1988 as part of the International Conference on Micro, Electro and Optomechanical Systems and Components (Madou 2011).

MEMS' potentials arose with microelectronics, as these complex devices became more sophisticated. Their materials and designs were enhanced in order to perform better on fixed tasks needed in the microelectronics industry (Madou 2011; Folch 2016; Saliterman 2006). As time and research advanced, biological applications were added to the domain of their use. Currently, the MEMS industry has several established milestones in the microfabrication technologies, such as micro-molding, photolithography, 3D structure assembly, among others (Borenstein 2008). Since MEMS could also contain mechanical components including cantilevers or membranes, they are ideal for fulfilling sensing (pressure and flow sensors) and/or actuation (optical-beam handling) tasks (Council 1998).

Biomedical or Biological Micro-Electro-Mechanical Systems, more commonly referred to as BioMEMS, are defined as micro or nano-scaled devices or systems that are used for processing, delivering, manipulating, or analyzing biological and chemical entities for biological or biomedical applications (Bashir 2004). Even though its name suggests the incorporation of both electronic and mechanical elements, it should be acknowledged that these devices do not necessarily integrate all functions in

R. Garcia-Ramirez · S. Hosseini (✉)
School of Engineering and Sciences, Tecnologico de Monterrey, Monterrey, Mexico
e-mail: samira.hosseini@tec.mx

S. Hosseini et al., *BioMEMS*, Lecture Notes in Bioengineering,
https://doi.org/10.1007/978-981-15-6382-9_1

every single device. Despite the fact that BioMEMS are a trend nowadays, the origin of micro systems used in life sciences, such as biology and/or neurology, are surprisingly old. Figure 1.1 shows a timeline with some of the most important BioMEMS milestones. Although many authors (Madou 2011; Folch 2016; Saliterman 2006) have reported the origins of BioMEMS in the late 1980s and early 1990s, preceding studies had already created BioMEMS technology without the use of such term. Table 1.1 demonstrates the first BioMEMS platforms in chronological order in great detail.

1.2 BioMEMS

1.2.1 Advantages of BioMEMS

BioMEMS offer various advantages over traditional methods that are worthwhile exploring, including a small device dimension and sample volume, portability, reliability in replication, high throughput performance, multifunctionality, possible automation, among others. The small device dimension provides obvious advantages as these devices possess a potential for miniaturization, whether *in-vivo* or *in-vitro*, including reduced manufacturing costs for devices as μTAS (micro-total-analysis systems) and LOC (lab-on-a-chip) devices. The smaller devices also benefit from the small sample size and reagents, in order to perform the same reaction that bulkier devices would need. The physical space they require and ease of portability is another advantage that BioMEMS have in contrast to their counterpart large pieces of lab equipment. Furthermore, BioMEMS devices provide multifunctionality that allows individual instruments to be integrated within one single device. This, in turn, facilitates automation, a feature that plays vital role in such devices. Fully integrated and automated devices can run the analysis with least human intervention. This is of great importance particularly when facing unknown or newly known dangerous illnesses. Considering the portability and the lightweight of such devices, they make great candidates for extreme point of care (EPOC) in remote and/or rural settings where there are no centralized laboratories. Equipment-free readouts, mass transfer of data, and/or readouts via smart phones and devices are the alternative analytical strategies linked to BioMEMS.

Nowadays, BioMEMS are one of the fastest growing fields in the world, because of the implications that it could create in multiple industries including the health sector, in particular in hospitals and healthcare facilities (Experts 2019). The current research regarding BioMEMS has increased at an accelerating rate. Since the first use of the term BioMEMS in the 1990s, there has been a constant increase in publications regarding this field. According to Clarivate Analytics the number of cites per year containing BioMEMS as a keyword has increased from less than 100 in the year 2000 to over 1600 in 2018. The global BioMEMS Market stood at 2.45 billion dollars in 2014 and predictions suggest it will grow above 25% by 2024, mostly due to the

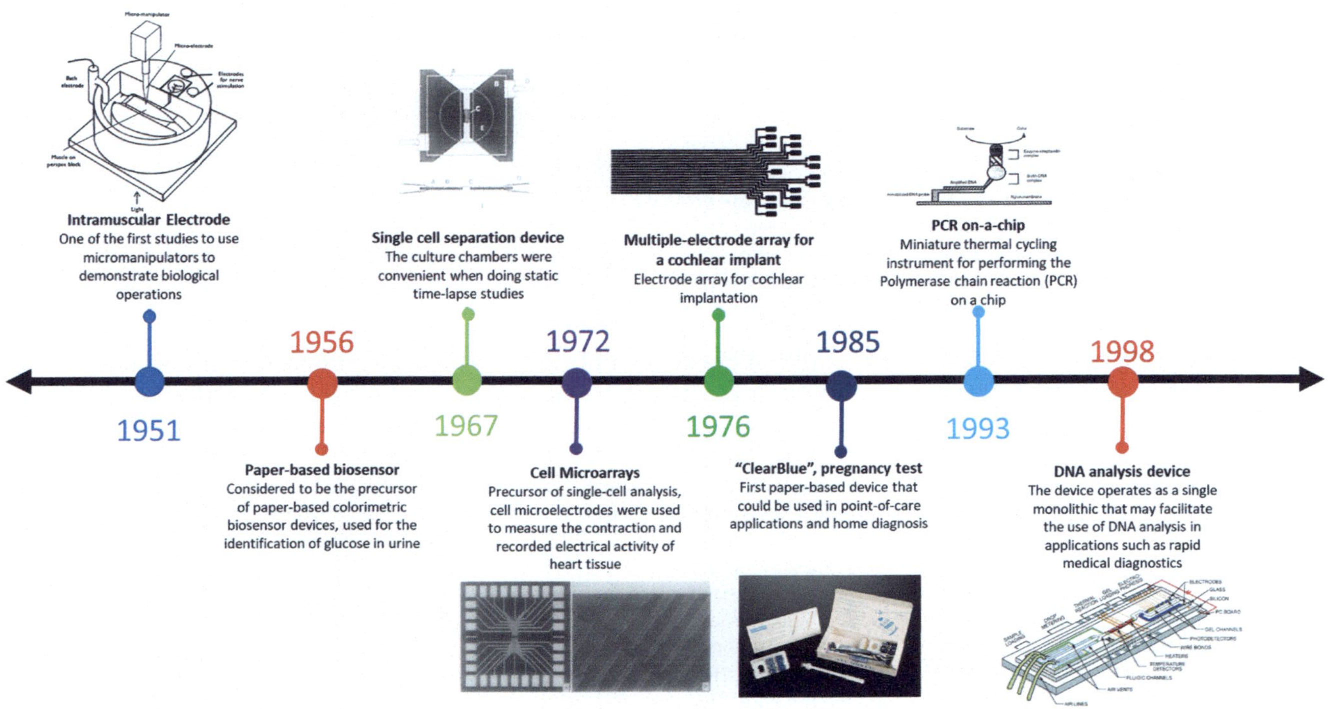

Fig. 1.1 Timeline of the History of BioMEMS including milestones in different areas of applications

Table 1.1 The first platforms in the history of BioMEMS

BioMEMS platform	Year	Scientist/authors/group	Company/institution	Country	Fabrication strategy	Application	Specific or remarks	References
Intramuscular microelectrode	1951	Paul Fatt and Bernhard Katz	University College London	England	Glass capillary tubing with external tip diameter of less than 0.5 μm were dipped into a beaker bridged to a calomel half-cell. Subsequently, the tips with electrodes were placed in a metal shield with a switch to connect the stimulator	Electrodes were used to measure the end-plate potential or local depolarization of muscle fibers	One of the first studies to use micromanipulators and microelectrodes to demonstrate biological operations	Fatt and Katz (1951)
Paper-based biosensor	1956	Alfred Free et al	Miles-Ames Research Laboratory Elkhart, Indiana	United States	A strip was impregnated with glucose oxidase, orthotolidine, and peroxidase	Simple test used for the identification of glucose in the urine based on a change in color	Considered to be the precursor of paper-based colorimetric biosensor devices	Free et al. (1956)
Single cell separation device	1967	Stephen Carter	Imperial Chemical Industries, Ltd.	England	A mask made by nickel electrodeposition was created, and palladium was deposited by evaporation onto a glass cover through the mask, building the desired culture chambers	The culture chambers allowed cells to attach to the surface of the device, and was proved useful for cell bonding, and single-cell studies	The culture chambers were convenient when doing static time-lapse studies, this is usually recorded as the first BioMEMS publication	Carter (1963)

(continued)

Table 1.1 (continued)

BioMEMS platform	Year	Scientist/authors/group	Company/institution	Country	Fabrication strategy	Application	Specific or remarks	References
Cell Microarrays	1972	C. A. Thomas Jr. et al	Harvard Medical School	United States	An acid-resistant, photosensitive polymer was used for coating glass. The platform was exposed to light via a photographic negative of each pattern. The platform was further developed and subsequently, a glass ring created within the insulated array using bees' wax in order to fix the position of the microelectrode tips	Cell microelectrode arrays were used to measure the contraction and recorded electrical activity of embryonic chick heart tissue	This device allowed multiple simultaneous recordings of electrical measurements in an array, the manufacturing process needs a special clean room facility and is costly	Thomas et al. (1972)
ISFET sensor for electrophysiology	1972	Piet Bergveld	Technische Hogeschool Twente	The Netherlands	Ion-sensitive field-effect transistor technology (ISFET) was manufactured with the help of the complementary metal–oxide–semiconductor (CMOS) technology and without any post processing steps	The sensor was intended to measure the ion activities in electrochemical and biological environments	The device measures ion activities without using a reference electrode, this limits the background noise and allows distinguishing series and parallel contribution of different ions	Bergveld (1972)

(continued)

Table 1.1 (continued)

BioMEMS platform	Year	Scientist/authors/group	Company/institution	Country	Fabrication strategy	Application	Specific or remarks	References
Implantable pressure sensing device	1974	Wilhelm Rindner	Device Research Inc.®	United States	Not disclosed	The device was designed to measure blood pressure within the body	The device enables the calibration of the pressure system while the sensor is at the site	Rindner (2020)
Multielectrode probe for neurophysiology	1975	Kensall Wise and James Angell	University of Michigan	United States	A gold electrode was formed by electroplating. Insulation material was removed from the electrode tip using photoengraving techniques	Multielectrode structure suitable for measuring neurons electrical activity with low noise. The three tips permitted the analysis of adjacent areas of the brain	The device is able to combine stimulating and recording electrodes in a single array while minimizing crosstalk. The low impedance of the electrodes reduces the overall noise of the system	Wise (1975)
Multiple-electrode array for a cochlear implant	1976	Graeme Clark and Richard Hallworth	University of Melbourne	Australia	A layer of platinum was deposited on a fluorinated ethylene-propylene (FEP) surface using a radio-frequency sputtering technique. The platinum layer was dip-coated in photoresist material, this was exposed to ultraviolet light (UV) through a photographic mask having the design of electrodes, therefore the unexposed areas were etched away	Electrode array for cochlear implantation	Reduction in the intracochlear section enables current density was kept at a minimum rate, thus not damaging cochlear tissue	Clark et al. (1977)

(continued)

Table 1.1 (continued)

BioMEMS platform	Year	Scientist/authors/group	Company/institution	Country	Fabrication strategy	Application	Specific or remarks	References
Microneedle for drug delivery	1976	Martin Gerstel and Virgil Place	Alza Corporation®	United States	The device was made by first mixing plastic pellets or powders and a drug, and then feeding the mixture into a heated barrel. It was conveyed by the thrust of a ram into a template which had the desired form of the microneedles	Drug delivery and therapeutics	First drug delivery system (DDS) for percutaneously administering drugs capable for the continuous administration of controlled amounts of drugs, eliminating the associated pain of conventional DDS	Gerstel and Place (1976)
Fixed-array multi-4electrode for neuronal monitoring	1977	Guenter Gross et al	Max Planck Institute for Psychiatry	Germany	The microelectrodes were produced by vacuum deposition of a titanium (Ti) and a gold (Au) film, this film was coated with a photo sensible polymer, then exposed to UV light through a photo-negative mask, the exposed polymer was removed via etching, and the unprotected metal was subsequently gold plated galvanically	Long-term monitoring of single unit extracellular electrical activity of neurons	The device provides low intrinsic electrode noise, with a versatile material combination to enhance temperature tolerance	Gross et al. (1977)

(continued)

Table 1.1 (continued)

BioMEMS platform	Year	Scientist/authors/group	Company/institution	Country	Fabrication strategy	Application	Specific or remarks	References
Biosensing gas chromatographic air analyzer	1979	Stephen Terry et al	Stanford University	United States	Microchannels were fabricated using photolithography and etching techniques, the microchannels were bonded with glass using heat and a voltage potential between the Si and the glass	Gas sensing for air contaminant analyzer	This device had a pocket-sized package, while closely retaining the performance of a larger device, it controlled the sequencing of the gas chromatography (GC) operation and analyzed the detector output signal simultaneously	Terry et al. (1979)
"Clearblue", pregnancy test	1985	Unipath Inc.®	Unipath Inc.®	United Kingdom	Not disclosed	Hand-held pregnancy testing	First paper-based device that could be used in point-of-care applications and home diagnosis, previous devices needed at least 30 min to give a result, while this version only needed 3 min	Jones and Kraft (2004)

(continued)

Table 1.1 (continued)

BioMEMS platform	Year	Scientist/authors/group	Company/institution	Country	Fabrication strategy	Application	Specific or remarks	References
Lab-on-a-chip for signaling	1987	Harvey Hoch et al	Comell University	United States	Polystyrene replicas of ion-etched Si wafer templated with specific surface topographies to exhibit maximum cell differentiation were microfabricated by using electron-beam lithography	Cell differentiation	The mechanisms involved in signal reception for the growth orientation and infection structure formation were determined for a stomatal penetrating rust fungus	Hoch et al. (1987)
Lab-on-a-chip for growth orientation	1987	Jochen Walter et al	Max Planck Institute for Developmental Biology	Germany	A negative mold is produced by a photoetching process, a positive mold is then formed; this is used for the production of the silicone rubber matrix containing the array of microchannels	Cell differentiation	Fast prototyping technique that enables the production of multiple devices from a single negative mold	Walter et al. (1987)
μTAS	1990	Andréas Manz et al	Ciba-Geigy®	Switzerland	Not disclosed	Potential uses for microfluidic and lab-on-chip devices	μTAS, created a novel strategy for further devices to explore the field	Manz et al. (1990)

(continued)

Table 1.1 (continued)

BioMEMS platform	Year	Scientist/authors/group	Company/institution	Country	Fabrication strategy	Application	Specific or remarks	References
PCR on-a-chip	1993	M.A. Northrup et al	Lawrence Livermore National Laboratory	United States	The device was manufactured on Si wafers and processed with silicon nitride (Si_3N_4) using low pressure chemical vapor deposition (LPCVD). Patterns were made by conventional photolithography techniques. Reaction chambers were bonded by depositing a thin film of low-temperature-curing polyimide between two wafers	Thermal cycling instrument for execution of the polymerase chain reaction (PCR) on a chip	The device shows significant improvements over commercial thermal cycling instrumentation including low power requirements, and short cycle times	Northrup et al. (1993)
Lab-on-a-chip for cell characterization	1993	Larry Kricka	University of Pennsylvania	United States	Silicon microchannels were fabricated by using selective etching, the microchannels were sealed with glass by diffusive bonding	Cell characterization	Same samples can be used subsequently in other procedures	Kricka et al. (1993)
Capillary electrophoresis for molecule separation	1994	Zhonghui Fan and Jed Harrison	University of Alberta	Canada	The device was fabricated using a modification of bulk Si micromachining methods, and standard photolithographic methods	The presented device integrated a capillary electrophoresis system on a chip capable of performing rapid separation of amino acids	The device competes with commercial chemical sensors, while the high thermal conductivity and mass of the glass substrate allows the use of higher electric fields	Fan and Harrison Jan. (1994)

(continued)

Table 1.1 (continued)

BioMEMS platform	Year	Scientist/authors/group	Company/institution	Country	Fabrication strategy	Application	Specific or remarks	References
Chambers for studying microtubule dynamics	1997	Timothy Holy et al	Princeton University and Bell Laboratories	United States	Chromium (Cr) was evaporated onto clean coverslips, photoresist was then exposed to UV through a mask and developed. The chromium was etched away in the exposed areas, and the glass was etched in buffered hydrofluoric acid (HF) to the desired depth	Cellular biology, study of diseases involved with cellular structure	The low cost of fabrication permits its fast prototyping and bulk production, nevertheless the production requires highly corrosive and toxic compounds	Holy et al. (1997)
Microneedles for drug delivery	1998	Sebastien Henry et al	Georgia Institute of technology	United States	A chromium masking material was deposited onto Si wafers and patterned. The wafers were then loaded into a reactive ion etcher	Painless drug delivery systems	Microneedles are long enough to cross the permeability barrier but short enough to avoid stimulating nerves	Henry (1998)
PDMS-based Microfluidic Systems	1998	David Duffy et al	Harvard University	United States	Master molds were fabricated via photolithography technique. The replica molds were made using glass posts placed on the master and casting PDMS to yield elastomeric replicas, to form enclosed channels. A PDMS replica was sealed to a flat slab of PDMS via plasma treatment	Inexpensive and easy BioMEMS device prototyping	The method was rapid and cost effective	Duffy et al. (1998)

(continued)

Table 1.1 (continued)

BioMEMS platform	Year	Scientist/authors/group	Company/institution	Country	Fabrication strategy	Application	Specific or remarks	References
DNA analysis device	1998	Mark Burns et al	University of Michigan	United States	All components were made using conventional photolithography and etching techniques. A glass substrate containing etched microfluidic channels was bonded to the Si substrate to seal the device. The finalized assembly was wire-bonded to a printed circuit board, and fitted with electrophoresis buffer wells	The device is capable of measuring DNA-containing solutions, mixing the solutions, amplifying or digesting the DNA, and separating and detecting those products	The device operates as a single monolithic system with a complex, low-power, integrated analysis systems at low unit cost	Burns et al. (1998)
Microfluidics with heterogeneous laminar flow	1999	Paul Kenis et al	Harvard University	United States	A wide capillary was assembled with PDMS, a two-electrode system was generated inside by flowing a three-phase laminar system of gold and water, a third electrode was created by depositing silver at the interface of the two phases to use it as a reference electrode	Fabricating more complex structures	Relatively complex method that allows for tailored devices	Kenis et al. (1999)

(continued)

Table 1.1 (continued)

BioMEMS platform	Year	Scientist/authors/group	Company/institution	Country	Fabrication strategy	Application	Specific or remarks	References
Lab-on-a-chip for patterning cells	1999	Shuichi Takayama et al	Georgia Tech	United States	A negative relief of PDMS was formed by curing a prepolymer on a salinized Si master having a positive relief of the capillary channels formed in photoresist on its surface	Selective treatment of cells	This study demonstrates the use of different types of cells adjacent to each other, the patterned delivery of chemicals to adhered cells, as well as performing enzymatic reactions over select cells or over a portion of a cell	Takayama (1999)

development of home care devices and hand-held analytical systems, followed by in-vitro diagnosis and biological and pharmaceutical research (Cooper 2016). Lower tendencies but important research fields are focused on disease-prediction devices to prevent fatal diseases including cancer, HIV, etc. (Experts 2019).

BioMEMS can be classified into two general groups based on their applications (Cima 2011): firstly, those created for biomedical purposes including inertial sensors or implants, and, secondly, those that embed microelectronics and micromachining techniques in order to acquire, sense, or manipulate biological or chemical entities (Folch 2016; Ramesham 2000).

1.2.2 First Attempts in Design and Fabrication of BioMEMS

BioMEMS devices related to biological entities most likely evolved from the need to analyze single cells *in-vitro* since previous techniques were mostly invasive, which resulted in damage to the cell and the inability to maintain long term connections (Thomas et al. 1972). For this reason, an effort to embed electrodes in the culture platform gave rise to the first microelectromechanical systems used in pair with biological agents. These first attempts date back to the late 1960s (Carter 1963) and continued to be replicated and improved until reliable results positioned MEMS as a viable technology for chronic non-invasive single cell studies (Regehr et al. 1988).

In October 1967 in England, Stephen Carter from the Imperial Chemical Industries, Ltd., was the first scientist to report a BioMEMS in his publication related to cell studies, where he described surfaces with patterned cellular adhesiveness using microtechnology. Carter coined the cell behavior term "haptotaxis" in 1965, when he used a method to create cellular patterns of mouse fibroblast onto palladium as mask (Regehr et al. 1988). He used a wire in contact with the acetate substrate that had created a smooth metallic gradient on the acetate, resulting in substrate-directed cell motion. Carter then repeated the shadow-evaporation technique in 1967 where the wire was a perforated 15 μm thick nickel stencil mask made by photochemical machining (Carter 1963). Additional information regarding the fabrication strategy of these devices are presented in Table 1.1. In his experiment, the prediction and the feasible fabrication of massively parallel single-cell assays was introduced for studying single-cell spreading confined into adhesive islands. This latter publication has marked the first report in the history of BioMEMS (Folch 2016; Saliterman 2006).

The first micro systems embedded in larger systems were developed in the disciplines of neuroscience and neurobiology (Urban 2007). In the first decades of the twentieth century, single glass sensors were used to measure physiological parameters on the surface of human tissues (Bates 1963). The miniaturization of these devices offered a wider comprehension of brain functions, information processing, and tissue abnormalities (Council 1998; Bashir 2004).

For clinical disciplines, including cardiology (Konstantinov et al. 2004), neurology (Bates 1963), and endoscopic surgery, many appliances were developed by means of delicate machinery and mechatronic technology long before the BioMEMS term was conceived. Endoscopic surgery had been performed in the early nineteenth century (Rathet et al. 1974). The development of miniaturization techniques, in combination with new materials and micro-optic units, led to the earliest use of BioMEMS technology. A fully implantable pacemaker was developed in 1958 (Konstantinov et al. 2004), and the first hybrid multi-channel cochlear implant was introduced in 1977 (Hochmair-Desoyer et al. 1983). Table 1.1 summarizes a number of biomedical devices and their main components along with their fabrications and main applications (Council 1998; Borenstein 2008; Experts 2019).

The first idea of a BioMEMS biosensor was proposed by Leland Clark and Champ Lyons, from the Medical College of Alabama, in 1962 (Clark and Lyons 1962). They proposed a quantification mechanism for glucose which included the action of an enzyme, a glucose oxidase, that would degrade glucose in the presence of oxygen and generate gluconic acid. This would create a change in pH that could be then detected by an electrode. The proposed strategy fully evolved to become a patent published in 1970 (Wang 2001). In 1976, the first BioMEMS microneedle for Drug Delivery System (DDS) was reported by Alza Corporation. The reported device consisted of a DDS for percutaneous administration of a drug comprising projections, a drug reservoir, and an extend of the projections for penetrating the stratum corneum (Gerstel and Place 1976).

MEMS technology was not introduced exclusively to the biology and medicine fields, but also in chemical applications. The first miniaturized ion-sensitive field-effect transistor sensor was invented in 1972 by Piet Bergveld. This device was produced by means of semiconductor and microfabrication techniques (Bergveld 1972). Furthermore, a gas chromatographic air analyzer fabricated on silicon (Si) wafer was developed by Terry et al. in 1979. The design and the techniques employed to manufacture this MEMS device were inspired by micromachining fabrication. This implementation for chemical sensing is another example of MEMS technology applied to chemical applications (Terry and Jerman 1979). Table 1.1 discusses further details regarding the fabrication strategy and the applications of both devices.

In 1985, Unipath Inc. (Bedford, United Kingdom) commercialized Clearblue, a pregnancy test that can be considered the first paper-based microfluidic device. Clearblue is a diagnostic home kit which relies on specific conjugation of antibodies to biomarkers for analysis of specific hormones associated with early pregnancy markers in human urine (Jones and Kraft 2004). Its capacity to give an instantaneous result is what made it different from previously developed and patented products. This system was based on the detection of Human Chorionic Gonadotropin (hCG) applying monoclonal antibodies attached to a Si solid phase, which created a coupled complex with a specific antigen present during the firsts weeks of pregnancy. When this complex of antigen–antibody was formed, the labeled immunoglobulins released a change of color over the solid phase. Therefore, this assay is considered as one of the first paper-based colorimetric analytic methods in the history of BioMEMS (Folch 2016).

As the field began to grow, different devices were invented and new terms had to be developed to define the distinct technologies. In 1990, A. Manz and H. Michael Widmer from Switzerland proposed the term micro total analysis system (μTAS) to present the idea of miniaturized total chemical analysis systems for chemical sensing (Manz et al. 1990). Seven years later, R.A. Lewis was the first person to use the term "lab-on-a-disc" to refer to a disc for genetic disease testing developed by Gamera Bioscience Corp (Lewis 1997).

In 1993, George Whitesides' group, introduced polydimethylsiloxane (PDMS) based microfabrication that revolutionized the BioMEMS field by introducing a cost-effective class of materials as desirable platforms for fabrication of micro devices (Duffy et al. 1998). The main advantages of using PDMS include a faster production rate compared to glass or Si microfluidic chips and cost-effectiveness compared to other production methods (Duffy et al. 1998). Moreover, PDMS is transparent at optical frequencies, which facilitates the observation of contents in microchannels in an easier way. It is considered biocompatible, and has low autofluorescence (Piruska et al. 2005). It is deformable, which allows the integration of other microfluidic components into the microchannels. The PDMS bonds to glass or another PDMS layer with a simple plasma treatment, allowing the creation of multilayers in a single microfluidic chip (Xia et al. 1999). Since the introduction of PDMS-based devices, a wide range of platforms made by soft-lithography techniques was used in microfluidics and is considered a cornerstone in multiple applications.

In 1994, Fan and Harrison, reported the first microchannels for molecular separation that introduced the possibility for separation of amino acids and molecular characterization (Fan and Harrison 1994). Other examples of microfluidic devices for cell-related applications emerged in the late 1990s, including the use of nucleic acid arrays, DNA analysis devices, and chambers for studying microtubule dynamics (O'Donnell-Maloney and Smith 1996; Holy et al. 1997; Burns et al. (1998).

In 1996 at University of Michigan, David Burke and his team researched on the movement of single nanoliter droplets though manipulating the water tension with heat, thus enabling the system to mix, to separate, and to measure volumes in a nano-scale. This system consisted of a polymerase chain reaction (PCR) preparation protocol including DNA enzymatic digestion, electrophoresis, and PCR confined in a single device (Burns et al. 1998). After this pioneering advancement, other single-droplet microfluidic mechanisms and devices were presented for various applications (Choi and Ng 2012).

1.2.3 General Applications of BioMEMS

BioMEMS can be used for a wide range of applications, including diagnostics, tissue engineering, analytical techniques, microfluidics, and biosensors, among others. The development of BioMEMS can help enhance the accuracy and precision of analysis in different applications. While the above mentioned are important areas of impact, there are many other applications for BioMEMS in chemistry, biotechnology, and the food industry. For example, biosensors can be used for rapid detection of foodborne pathogens, such as E. Coli and Salmonella, which are a major problem when importing foreign food into a country (Yoon and Kim 2012). This comprehensive book, however, solely focuses on the latest updates on BioMEMS in their major biosensing applications. Figure 1.2 shows some of the major applications of BioMEMS and an example for each field (Fig. 1.3).

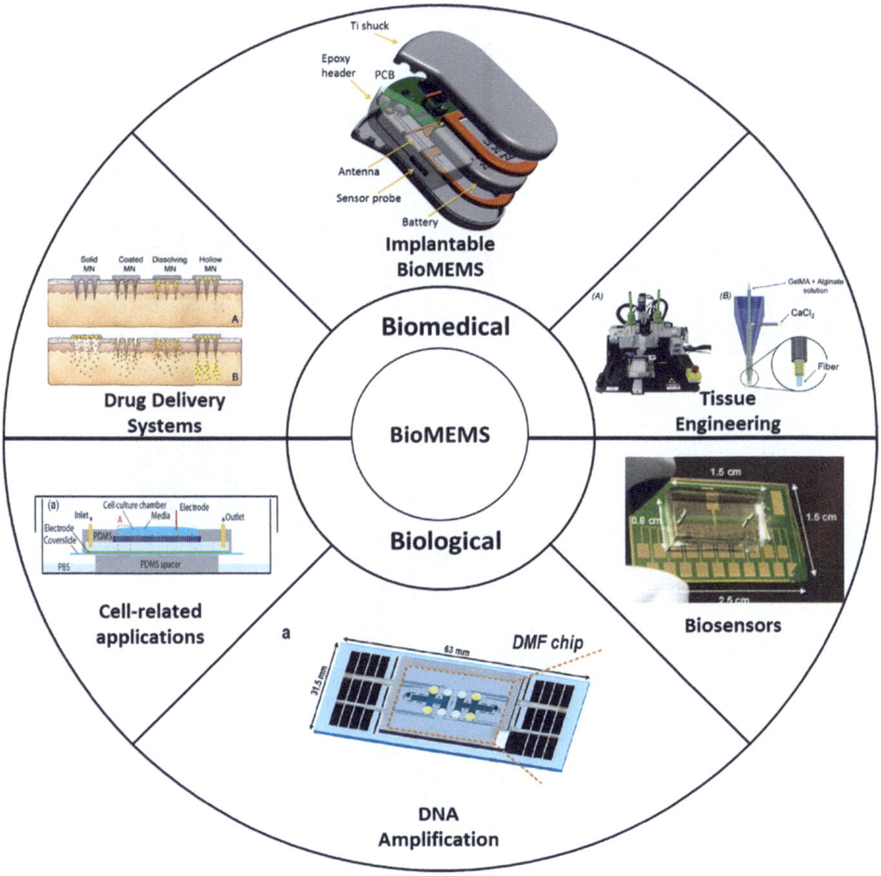

Fig. 1.2 Different areas of application of BioMEMS

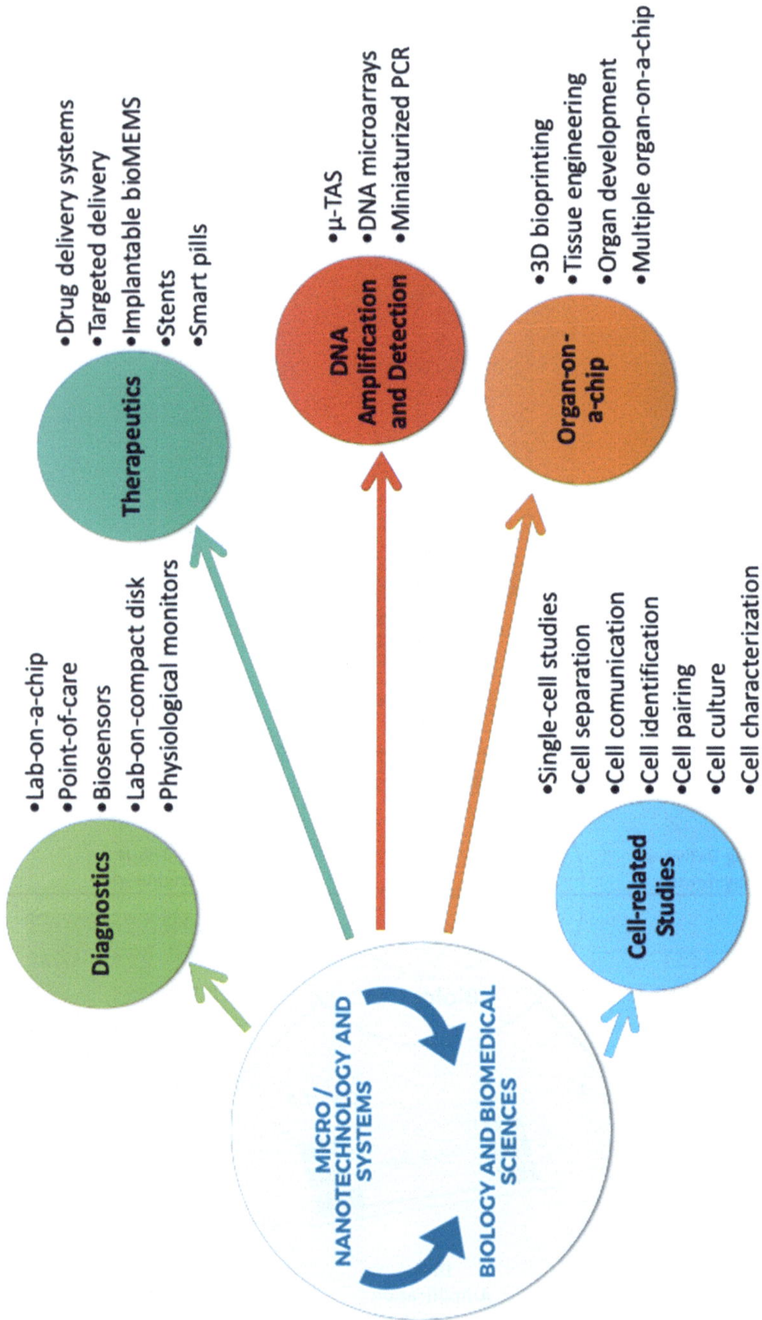

Fig. 1.3 Applications of BioMEMS in different areas

References

Bashir R (2004) BioMEMS: state-of-the-art in detection, opportunities and prospects. Elsevier
Bates JA (1963) Special investigation techniques: indwelling electrodes and electrocorticography
Bergveld P (1972) Development, operation, and application of the ion-sensitive field-effect transistor as a tool for electrophysiology. https://ieeexplore.ieee.org/Xplore/home.jsp
Borenstein JT (2008) BioMEMS technologies for regenerative medicine. https://www.cambridge.org/
Burns MA et al (1998) An integrated nanoliter DNA analysis device
Carter SB (1963) Haptotactic islands a method of confining single cells to study individual cell reactions and clone formation
Choi K, Ng A (2012) Digital microfluidics. https://www.annualreviews.org/
Cima MJ (Jul.) Microsystem technologies for medical applications. Annu Rev Chem Biomol Eng 2(1):355–378. https://doi.org/10.1146/annurev-chembioeng-061010-114120
Clark LC, Lyons C (1962) Electrode systems for continuous monitoring in cardiovascular surgery. Ann NY Acad Sci 102(1):29–45. https://doi.org/10.1111/j.1749-6632.1962.tb13623.x
Clark GM, Black R, Dewhurst DJ, Forster IC, Patrick JF, Tong YC (1977) A multiple-electrode hearing prosthesis for cochlear implantation in deaf patients. Springer-Verlag
Cooper S (2016) BioMEMS market analysis, market size, application analysis, regional outlook, competitive strategies and forecasts, 2016–2024. Hexa Res: 90
Council NR (1998) Microelectromechanical systems: advanced materials and fabrication methods
Duffy DC, Mcdonald JC, Schueller OJ, Whitesides GM (1998) Rapid prototyping of microfluidic systems in poly(dimethylsiloxane). https://doi.org/10.1021/ac980656z
Experts I (2019) Global BioMEMS devices market overview 2016–2022—home care devices. In: Vitro diagnostics, medical devices and pharmaceutical and biological research
Fan ZH, Harrison DJ (1994) Micromachining of capillary electrophoresis injectors and separators on glass chips and evaluation of flow at capillary intersections. Anal Chem 66(1):177–184. https://doi.org/10.1021/ac00073a029
Fatt P, Katz B (1951) An analysis of the end-plate potential recorded with an intra-cellular electrode
Folch A (2016) Introduction to bioMEMS
Free AH et al. (1956) Simple specific test for urine glucose
Gerstel MS, Place VA (1976) United States Paten Gerstel et al. (19) (54) 75 Drug delivery device
Gross G, Rieske E, Kreutzberg G (1977) A new fixed-array multi-microelectrode system designed for long-term monitoring of extracellular single unit neuronal activity in vitro. Elsevier
Henry S et al (1998) Microfabricated microneedles: a novel approach to transdermal drug delivery. J Pharm Sci 87(8):922–925
I. Hochmair-Desoyer, Hochmair ES, Burian K, Stiglbrunner HK (1983) Percepts from the Vienna cochlear prosthesis. https://psycnet.apa.org/
Hoch, HC, Staples RC, Whitehead B, Comeau J, Wolf ED (1987) Signaling for growth orientation and cell differentiation by surface topography in Uromyces. https://science.sciencemag.org/
Holy TE, Dogterom M, Yurke B, Leibler S (1997) Assembly and positioning of microtubule asters in microfabricated chambers. Natl Acad Sci 94(12):6228:6231
Jones G, Kraft A (2004) Corporate venturing: The origins of unilever's pregnancy test. Bus Hist 46(1):100–122. https://doi.org/10.1080/00076790412331270139
Kenis PJA, Ismagilov RF, Whitesides (1999) GM Microfabrication inside capillaries using multiphase laminar flow patterning
Konstantinov I, Alexi-Meskishvili, VV, Williams WG, Freedom RM, Van Praagh R (2004) Atrial switch operation: past, present, and future. Elsevier
Kricka LJ, Nozaki O, Heyner S, Garside WT, Wilding P (1993) Applications of a microfabricated device for evaluating sperm function. https://academic.oup.com/journals
Lewis RA (1997) Gamera Bioscience Corp. develops lab on a disc for genetic disease testing. Mary Ann Liebert Inc. Publ 2
Madou M (2011) Manufacturing techniques for microfabrication and nanotechnology.

Maluf N, Williams K (2004) Introduction to microelectromechanical systems engineering
Manz A, Graber N, Widmer HM (1990) Miniaturized total chemical analysis systems: a novel concept for chemical sensing (1990)
Northrup M, Ching M, White R (1993) A memsbased miniature DNA analysis system, in transducers 93
O'Donnell-Maloney M, Smith C (1996) The development of microfabricated arrays for DNA sequencing and analysis. Trends Biotech 14(10):401:407
Piruska A, Nikcevic I, Lee S, Ahn C (2005) The autofluorescence of plastic materials and chips measured under laser irradiation. https://pubs.rsc.org/
Rindner W (2020) Pressure sensing device. 1974, Google Patents. Google Académico [Online]. Available https://scholar.google.com/scholar?hl=es&as_sdt=0%2C5&q=Rindner%2C+W.%2C+Pressure+sensing+device.+1974%2C+Google+Patents.&btnG. Accessed 31 Jan 2020
Ramesham R (2000) Fundamentals of microelectromechanical systems (MEMS)
Rathet P, Lutzeyerm W, Godwin W (1974) Philip Bossini and the Lichtleir. Urology 3:113–123
Regehr WG, Pine J, Rutledge DB (1988) A long-term in vitro silicon-based microelectrode-neuron connection (1988)
Saliterman S (2006) Fundamentals of BioMEMS and medical microdevices
Takayama S, McDonald JC, Ostuni E, Liang MN, Kenis PJA, Ismagilov RF, Whitesides GM Patterning cells and their environments using multiple laminar fluid flows in capillary networks. Natl. Acad Sci
Terry S, Jerman J (1979) A gas chromatographic air analyzer fabricated on a silicon wafer. https://ieeexplore.ieee.org/Xplore/home.jsp
Thomas Jr CA, Springer PA, Loeb GE, Berwald-Netter Y, Okun LM (1972) A miniature microelectrode array to monitor the bioelectric activity of cultured cells. Elsevier
Urban G (2007) BioMEMS 16. Springer Science. Google Académico [Online]. Available https://scholar.google.com/scholar?hl=es&as_sdt=0%2C5&q=Urban%2C+G.%2C+BioMEMS.+Vol.+16.+2007%3A+Springer+Science+%26+Business+Media.&btnG. Accessed 30 Jan 2020
Walter J, Kern-Veits B, Huf J, Stolze B, Bonhoeffer F (1987) Recognition of position-specific properties of tectai cell membranes by retinal axons in vitro
Wang J (2001) Glucose biosensors: 40 years of advances and challenges. Electroanal Int J Devoted Fund Pract Aspects Electroanal 13(12): 983–988
Wise K (1975) A low-capacitance multielectrode probe for use in extracellular neurophysiology. https://ieeexplore.ieee.org/Xplore/home.jsp
Xia Y, Rogers JA, Paul KE, Whitesides GM (1999) Unconventional methods for fabricating and patterning nanostructures. Chem Rev 7:1823–1848. https://doi.org/10.1021/cr980002q
Yoon J-Y, Kim B (2012) Lab-on-a-chip pathogen sensors for food safety. Sensors 12:10713–10741. https://doi.org/10.3390/s120810713

Chapter 2
Bio-microelectromechanical Systems (BioMEMS) in Bio-sensing Applications-Colorimetric Detection Strategies

Michelle Alejandra Espinosa-Hernandez, Sofia Reveles-Huizar, and Samira Hosseini

2.1 Introduction

One of the most well-known analytic methods of detection is based on quantifying the UV–vis absorption or reflection, also known as colorimetric (Li et al. 2019a). Colorimetric biosensors are simple, rapid, disposable, and low-cost (Zhang et al. 2006; Li et al. 2019b). Additionally, they enable the signal readout with color changes or even directly judged by the naked eye (Li et al. 2019b). For MEMS, specially BioMEMS, these qualities enable medical exams to become more accessible and user friendly. Likewise, biological processes can be carefully assessed due to the concentration of the color which is commonly directly proportional to the concentration of the analyte of interest (Aldewachi et al. 2018). In this chapter, we review the latest advancement of paper-based, microfluidics and alternative BioMEMS that operate based upon the principles of colorimetric detection. A comprehensive review of these devices is presented and summarized in Table 2.1.

2.2 Colorimetric Detection Strategy

The detection method known as colorimetric determines the presence and concentration of an analyte through comparing the color changes of a solution. This method is based on the quantification of the change in the absorbance of the solution due to the concentration of the analyte of interest. This relationship can be described using the Beer-Lambert equation, which relates the concentration to the absorption, with the light path remaining constant (Wilson 2013). When white light passes through a

M. A. Espinosa-Hernandez · S. Reveles-Huizar · S. Hosseini (✉)
School of Engineering and Sciences, Tecnologico de Monterrey, Monterrey, Mexico
e-mail: samira.hosseini@tec.mx

S. Hosseini et al., *BioMEMS*, Lecture Notes in Bioengineering,
https://doi.org/10.1007/978-981-15-6382-9_2

Table 2.1 Recent BioMEMS platforms for colorimetric detection: Type of the platform, main components, fabrication strategy, mechanism of operation. The analyte of interest, and the advantages and disadvantages of each platform are presented in this table

BioMEMS platform	Main components	Fabrication strategy	Mechanisms of operation	Detected analyte	Specifics	References
PAD DNA sensor based on acpcPNA-induced nanoparticle aggregation	• AgNP • acpcPN probe • PAD Multiplex DNA • Sensor	The PAD was made with a wax-printing technique and the sensor by folding two-layer halves while a PDMS lid held it together	The sample reservoir was punched through the whole device. The acpcPNA probe and AgNPs solution were added onto the detection and control zones. Finally, the sample solution was added onto the sample reservoir and flow through the channels to wet the colorimetric detection zones	• MERS • TB • MERS-CoV • HPV	This colorimetric DNA sensor exhibited high selectivity and has the potential to be a low-cost and disposable alternative tool for rapid and selective screening and detecting in infectious diseases	Teengam et al. (2017)

(continued)

Table 2.1 (continued)

BioMEMS platform	Main components	Fabrication strategy	Mechanisms of operation	Detected analyte	Specifics	References
PAD ELISA platform	• Wax-printed 96-microzone paper plate • Image interpretation device • AuNPs	The AuNPs were synthesized through citrate-reduction techniques, later joined with PEG and were added to the IgG antibodies. A wax printing method was used to create the pattern on the paper designed. The dimensions were according to the standard Costar 96-well microtiter plate	A direct ELISA protocol was performed in the device to detect the IgG with a colorimetric substrate. Multiple P-ELISAs were conducted in parallel, allowing the use of common microplate processing techniques with the P-ELISA format	Neuropeptide Y	This device provides an inexpensive platform to carry out highly sensitive (picomolar level) biomolecular assays at very low sample and reagent volumes	Murdock et al. (2013)

(continued)

Table 2.1 (continued)

BioMEMS platform	Main components	Fabrication strategy	Mechanisms of operation	Detected analyte	Specifics	References
Integrated PAD biosensor incorporating nucleic acid extraction	• Silica microbead channels • Siphon channels • Glass fiber • Absorbent pads • Rotary machine • Heating block	The device is composed of three major microfluidic layers joined by double sided adhesive film. The micropatterns of the first and third layers were fabricated on polycarbonate sheets, and the microstructures on the second layer were fabricated on a PMMA sheet using a CNC milling machine. The lateral flow strips were inserted between the second and the third layer	The lysate mixture was introduced into the inlet, which was filled in a channel by capillary forces. The lysate debris in the microbead-bed channel flowed out into the waste chamber, followed by the washing buffer eruption. The solution passed through the microbeads to carry the adsorbed DNA on the microbeads into the LAMP chamber and the LAMP reaction mixture was also transferred into the LAMP chamber. Finally, the LAMP products and the running buffer solution were loaded onto the lateral flow strip via the connecting channels	• Streptococcus pneumonia • E. coli	This device can perform simple nucleic acid extraction, amplification and colorimetric detection by the naked eye in about an hour, thus making it a high performance microdevice	Choi (2016)

(continued)

Table 2.1 (continued)

BioMEMS platform	Main components	Fabrication strategy	Mechanisms of operation	Detected analyte	Specifics	References
PAD smartphone accessory for biomarkers	• Smartphone application • Test strips • Smartphone case with inlet slot • Smartphone	The indicator strip consisted of a cutout of a pHydrion Spectral plastic pH indicator strip for sweat testing and a strip for saliva testing. The reference strip was made of white plastic material. The flash diffuser consists of a PDMS membrane. The smartphone case was 3D printed in order to isolate the test strip from external light	The app was loaded, and the strip was selected for the readout. The calibration data updated while the sample was inserted	pH in saliva and sweat	This device allowed noninvasive real-time analysis by means of disposable test strips	Oncescu et al. (2013)

(continued)

Table 2.1 (continued)

BioMEMS platform	Main components	Fabrication strategy	Mechanisms of operation	Detected analyte	Specifics	References
Glass nanofibers colorimetric biosensor	• Detection ribbon made from hardened PVC • Detection fabric with immobilized and stabilized AChE • Cellulose paper strip impregnated with ATChI and Ellman's reagent	The biosensor was a plastic strip with an indication fabric and the carrier impregnated with a substrate and an indicator. The Ellman's reagent is used as a color indicator, joined with the substrate on the cellulose filter paper. To enhance the color, glass nanofibers were used as the substrate	The moistened detection fabric was exposed to air, contaminated water or pressing a wet detection zone to the test surface resulting in color change. Yellow was the indicator that no inhibitors exist while white demonstrated the presence of inhibitors	Cholinesterase inhibitors	The paper made of glass nanofibers provided a better color effect than the standard cellulose paper within the same 2 min	Matějovský and Pitschmann (2018)

(continued)

Table 2.1 (continued)

BioMEMS platform	Main components	Fabrication strategy	Mechanisms of operation	Detected analyte	Specifics	References
Paper/polymer hybrid μPAD microplate	• PMMA • Chromatography paper disks • Microwells	Micro-machining using laser ablation was used. Pieces of chromatography paper were cut using a laser cutter and placed inside the device. To assemble the device, different PMMA layers were clamped together	Assay reagents were loaded in the top PMMA layer. Then, the reagents arrived to the chromatography paper in each microwell by capillary forces and allowed the excess of it flow to the third layer for removal	• HBsAg • IgG	The use of paper-based technology resulted in rapid immobilization of antibody/ antigen and avoided complicated surface modifications. The device was simple, portable, rapid and highly sensitive	Sanjay et al. (2016)
Double-layered μPAD	• Nylon mesh • Whatman paper	The 3D microfluidic channels were made through traditional wax-screen printing technology	A blood sample was dispensed on sampling zone, and passed through the hydrophobic channels in order to react with the reactants. The color intensity was recorded by an Image J software	• Glucose • Uric acid • Lactate • Choline	This device allowed the detection of four biomolecules simultaneously, and improved colorimetric performance, sensitivity, and detection range	Li et al. (2018)

(continued)

Table 2.1 (continued)

BioMEMS platform	Main components	Fabrication strategy	Mechanisms of operation	Detected analyte	Specifics	References
Multilayer μPAD coupled with filtration	• Chromatography paper • Filter paper	Channels were printed using a wax printer. The bottom layer was made of Whatman paper. All the layers were inserted in a hydrophobic wax paper holder	Colorant reagents were applied to the channels to form metal hydroxides. These ions captured the antibiotics and created metal complexes which could not go through the filter paper. The rest of the liquid was absorbed for colored complexes to be seen	• Oxytetracycline • Norfloxacin	With a wider range of detection agents, greater recovery rate, and quicker assembly and testing, this device and method was found to be a valuable platform for food safety surveillance	Nilghaz and Lu (2019)
Colorimetric μPAD Biosensor	• Paper substrates • Wax printer • Adhesive tape	The biosensor was wax-printed on paper platforms and modified with chitosan as explained previously	Tears were placed on the inlet area and pulled up to the detection zone by capillary forces. The color changes were detected by an office scanner and converted to Red–Green–Blue scale	GLU	The device was proposed as a noninvasive alternative capable of distinguishing different concentration levels in a sample to control glucose for diabetic patients	Gabriel et al. (2017)

(continued)

Table 2.1 (continued)

BioMEMS platform	Main components	Fabrication strategy	Mechanisms of operation	Detected analyte	Specifics	References
Compact embeddable μCAD	• Cotton fabrics/cloth • Dual inlet	Wax patterning technique was utilized to pattern microfluidic channels on scoured-cotton cloth fabric and the 3D colorimetric microfluidic device was made by folding a 2D pattern	Solutions of the respective colorimetric reagents were pipetted into the detection zones. The inlet point was immersed into the urine. The sample analyte was introduced through an inlet point to react with the reagents	• GLU • Nitrite • Protein	The easy to pattern, inexpensive and environmentally friendly device was stable for a week proving useful for applications in underdeveloped areas	Nilghaz et al. (2015)
C-μPAD	Chromatography paper	The hydrophobic barriers were made by CVD of TCS on a chromatography paper	A pattern was cut out onto a reactive surface that was glued to paper surface and placed inside the CD chamber. A colorimetric reagent was immobilized for a stationary and uniform reaction through thermal condensation coupling method	• GLU • Metals	More complex detection chambers and higher degree of reliability resulted from this device by controlling temperature, pattern size, and CVD duration	Lam et al. (2017)

(continued)

Table 2.1 (continued)

BioMEMS platform	Main components	Fabrication strategy	Mechanisms of operation	Detected analyte	Specifics	References
Microfluidic colorimetric biosensor	• 8 microchannels • $CuFe_2O_4$/GQDs MNPs • Microfluidic chip	The microfluidic chip was made by traditional soft lithography technique. The colorimetric assay was conducted with peroxidase-like $CuFe_2O_4$/GQDs MNPs	Analytes of different concentrations were injected into the microchannel from two entrances simultaneously, and converged at the intersection point. ACh and NaH_2PO_4 buffer solutions were injected to other two entrances and were directed to the color area	• Chlorpyrifos inhibitors	The GQDs MNPs were added to enhance the colorimetric result. The device was compact and precise using small sample volume, lowering the limit of detection	Mao et al. (2017)

(continued)

Table 2.1 (continued)

BioMEMS platform	Main components	Fabrication strategy	Mechanisms of operation	Detected analyte	Specifics	References
Microfluidic colorimetric biosensor using AuNPs aggregation and smartphone imaging	• AuNPs • Microfluidic chip • 2 serpentine mixing channels • Separation chamber • Detection chamber	Solidworks was used to create the design of the chip. It was 3D printed and underwent surface plasma bonding. Surplus support material was removed while the PDMS prepolymer attached to it. The PDMS replica was peeled off and joined to the glass slide	The MNPs and the PSs reacted with the target bacteria in the first mixing channel. In the second step, AuNPs and the crosslinking agents reacted with the catalysate. This interaction was responsible for the color change	• E. coli O157:H7	The addition of AuNPs effectively increased the signal indication. Also, the chip's mixing productivity impacted the detection time and sensitivity for foodborne pathogens	Zheng et al. (2019)

(continued)

Table 2.1 (continued)

BioMEMS platform	Main components	Fabrication strategy	Mechanisms of operation	Detected analyte	Specifics	References
Smartphone-assisted microfluidic chemistry analyzer	• Microfluidic chip • Smartphone • Step motor • Microcontroller • Bluetooth module	The main structure was made by plastic injection molding and two PSA layers were added on either side. It included four reaction chambers for the multiplex measurements	Reaction mixture were pipetted into each detection chamber and allowed to dry, leaving one chamber empty for quality control. The LED emitted and transmitted by the reagent. The step motor rotated the camera through all four chambers and were then processed by the microcontroller. The color changes were measured by a custom analyzer and built-in optical system	• GLU • TG • TC	Multi-index monitoring was done with a high accuracy and low cost, while maintaining excellent consistency with conventional chemistry analyzers	Li et al. (2019c)

(continued)

Table 2.1 (continued)

BioMEMS platform	Main components	Fabrication strategy	Mechanisms of operation	Detected analyte	Specifics	References
Molecular diagnostic quantitative pregnancy test on a DVD	• DVD • Computer optical drive	The DVD's PS surface underwent activation by UV irradiation and treatment with EDC and NHS. A PDMS plate was added with microchannels. Anti-hCGα were immobilized on the surface, streptavidin nanogold conjugates were bound and a silver staining treatment was performed	Samples were loaded into the PDMS microfluidic channels on the DVD. The DVD was spun, and the samples create a different radial distance depending on the concentration of the analyte. The assay was tested with a diagnostic software and evaluated based on radial distance and optical darkness ratio	• hCG	Th strategy provided a low-cost POC tool for worldwide testing and comparable results to common immunoassays and ELISA	Li et al. (2014)

(continued)

Table 2.1 (continued)

BioMEMS platform	Main components	Fabrication strategy	Mechanisms of operation	Detected analyte	Specifics	References
LAMP Integrated rotary microfluidic system	• Centrifugal microfluidic platform • Glass microbeads solid phase matrix base	Multiple microfluidic layers were created using a CNC milling machine. To assemble the three layers, a PSA film was used after cutting the micropattern by a plotter. All layers were aligned and bonded by a hot press	This microdevice incorporates the whole procedure for nucleic acid-based test, including the solid-phase nucleic acid extraction, the LAMP reaction, and the lateral flow strip based colorimetric detection	• Salmonella Typhimurium • Salmonella • Typhimurium • Vibrio parahaemolyticus	High performances of the microdevice provided great potential as a user-friendly POC analyzer that can be applicable to resource limited environments	Park et al. (2016)
Smartphone spectrometer	• Smartphone • Grated CD • Built-in LED • CMOS camera	The grating tracks of the CD were tilted with respect to the LED but normal to the incident light so the light may be refracted from the CD on the camera. The flashlight was emitted through a pinhole in front of the LED	A bi-enzymatic cascade assay and peptide-functionalized AuNPs were used to detect glucose and troponin I respectively	• GLU • Troponin I	This device allows real-time measurements with LOD of approx. 50 ng mL-1. It presented similar results to commercial devices while offering a compact cost-effective tool	Wang et al. (2016)

(continued)

Table 2.1 (continued)

BioMEMS platform	Main components	Fabrication strategy	Mechanisms of operation	Detected analyte	Specifics	References
Electronic-based ELISA	• Circuit board • TFT touch screen • Au surface • Commercially available IFNγ	Using heat and adhesive properties, Cu foil was laminated on FR4 PCB substrate and a gold layer was plated on top. The PMMA wells were fixed on the PCB surface	TMB was used as the reporter reagent. Amperometric detection was performed by a second generation amperometry	• IFNγ	This device facilitated colorimetric and amperometric assays in an inexpensive and portable manner	Evans (2017)
Centrifugal microfluidic device integrated with LAMP	• PMMA layers • PSA layer • Loading and mixing chambers • Smartphone	The device was made of two PMMA layers and a middle PSA layer press-bounded together. A square-wave microchannel, metering chambers, and RPM control allowed the process to flow	The LAMP reagents and primers were injected to the loading chambers while the sealing material was loaded into the sealing chambers and the wax into the wax valve and the DNA samples were injected into the amplification chambers	• E. coli • Vibrio cholerae	This device allowed 30 simultaneous genetic analyses of three different foodborne pathogens. This process was performed in approx. an hour and presented a considerably low LOD	Sayad et al. (2017)

(continued)

Table 2.1 (continued)

BioMEMS platform	Main components	Fabrication strategy	Mechanisms of operation	Detected analyte	Specifics	References
RCA-based biosensor for attomolar detection	• Polymerase • MB amplification template • MB/GDNA probe • Spectrophotometer • Electrophoresis analyzer	N/A	miRNA triggered MB mediated strand displacement to release nicking triggers cyclically, leading to a TIRCA. This result was vast production of GDNA, which joining with hemin, form an HRP mimic, has catalyzed the colorimetric reaction		The machine united the advantages of enzymatic signal amplification and TIRCA. It allowed a decrease in background noise and improvement in stringent target recognition as well as sensitivity	Li (2016)

Paper-based analytical device (PAD); Pyrrolidinyl peptide nucleic acid (acpcPNA); Silver nanoparticles (AgNPs); Polydimethylsiloxane (PDMS); Middle East Respiratory Syndrome (MERS); Tuberculosis (TB); Middle East Respiratory Syndrome coronavirus (MERS-CoV); Human Papillomavirus (HPV); Enzyme-linked immunosorbent assays (ELISA); Gold nanoparticles (AuNPs); Poly(ethylene glycol) (PEG); Immunoglobulin G (IgG); Paper-based enzyme-linked immunosorbent assay (P-ELISA); Poly(methyl methacrylate) (PMMA); Loop-mediated isothermal amplification (LAMP); Escherichia coli (E. coli); Polyvinylchloride (PVC); Acetylcholinesterase (AChE); Acetylthiocholine iodide (ATChI); Microfluidic paper-based analytical device (μPAD); Hepatitis B Surface Antigen (HBsAg); 3,3′,5,5′-tetramethylbenzidine (TMB); Microfluidic cloth-based analytical device (μCAD); Chemically patterned μPAD (C-μPAD); Chemical vapor deposition (CVD); Trichlorosilane (TCS); Graphene Quantum Dots (GDQs); Magnetic nanoparticles (MNPs); Acetylcholine (ACh); Polystyrene (PS); Glucose (GLU); Triglycerides (TG); Total cholesterol (TC); 1-ethyl-3-(3-dimethylaminopropyl)carbodiimide (EDC); N-hydroxysuccinimide (NHS); Anti-human Chorionic Gonadotropin Gα (anti-hCGα); Human Chorionic Gonadotropin (hCG); Polycarbonate (PC); Point of care (POC); Pressure sensitive adhesive (PSA); Complementary metal oxide semiconductor (CMOS); Lab-on-a-Printed Circuit Board (LoPCB); Interferon Gamma (IFNγ); Copper (Cu); Printed Circuit Board (PCB); Pressure sensitive adhesive (PSA); Revulsion per minute (RPM); Limit of detection (LOD); Rolling circle amplification (RCA); Molecular beacon (MB); G-rich DNA (GDNA); MicroRNA (miRNA); Toehold initiated rolling circle amplification (TIRCA); Horseradish peroxidase (HRP)

colored substance, light wavelengths are absorbed in different proportions (Fuwa and Vallee 1963). The intensity of the resultant color will be proportional to the concentration of the measured analyte and the amount of absorbed light will be proportional to the intensity of the color (Ricci et al. 1994).

Colorimetric analysis is applicable to detect the presence of organic and inorganic compounds, making it a suitable option for biosensors. Some applications for colorimetric detection devices include hand-held bio-diagnostics, point-of-care diagnostics, and naked-eye detection. Current chapter focuses on the fabrication of microfluidic, paper-based, or polymer-based platforms based on this common detection strategy.

2.3 Recent Advances of Colorimetric Detection in Paper-Based BioMEMS

One of the major problems in healthcare nowadays is accessibility. Many people worldwide have limited to no access to laboratories or hospitals, hence, in order to reach these communities, smaller portable devices with the same accuracy are needed. Among these developments, the paper-based analytical devices (PAD) attract a great deal of attention. PAD devices have proven to be the inexpensive, simple, portable, and disposable. Likewise, they are easy to use, make complicated readout equipment unnecessary, and produce semi quantitative results (Teengam et al. 2017) in a short amount of time (Murdock et al. 2013). For that reason, PADs are being used to diagnose diseases via DNA and/or RNA recognition (Teengam et al. 2017), monitor human activity (Murdock et al. 2013), detect nucleic acids (Choi 2016), sense pH in sweat and/or saliva (Oncescu et al. 2013), and recognize cholinesterase inhibitors (Matějovský and Pitschmann 2018), thus, making such devices favorable for a wide range of applications including medical (Teengam et al. 2017), military (Murdock et al. 2013), nutrition (Choi 2016), biochemical (Oncescu et al. 2013), and nerve chemical warfare (Matějovský and Pitschmann 2018). Below, several examples of PADs are presented with a specific focus on biosensing application.

In addition to the previously stated benefits of paper-based devices, μPADs enhance point of care for detecting diseases (Sanjay et al. 2016), biomolecules (Li et al. 2018; Gabriel et al. 2017), and antibiotics (Nilghaz and Lu 2019). They offer inexpensive, simple, eco-friendly, portable and quick bioanalysis (Nilghaz and Lu 2019). Additionally, due to their small size, these devices provide a greater surface to volume ratio, improving the immobilization of proteins through processes such as enzymelinked immunosorbent assay (ELISA) integrated into this platform, and other biological agents (Sanjay et al. 2016). Many variations and additions can be made to μPADs in order to enhance its properties. For example, cotton, being a similar material to paper and providing the same advantages as well as being stronger and more durable, becomes an option for embedding into daily wearable products (Nilghaz et al. 2015). Also, by using chemical vapor deposition (CVD) instead of wax printing, more complex detection chambers and a higher degree of reliability for colorimetric detection result can be achieved (Lam et al. 2017). Some of the latest examples of the paper-based BioMEMS used for colorimetric detection are provided here.

Teengam et al. (2017) produced a paper-based colorimetric assay for DNA detection based on pyrrolidinyl peptide nucleic acid-induced nanoparticle aggregation in order to have a simple and quantitative means of detecting diseases such as Middle East Respiratory Syndrome (MERS), Tuberculosis (TB), Middle East Respiratory Syndrome coronavirus (MERS-CoV), and Human Papillomavirus (HPV). To create the device, a multiplex colorimetric PAD with a derived backbone from D-proline/2-aminocyclopentanecarboxylic acid (acpcPNA), silver nanoparticles (AgNPs) and a paper-based multiplex DNA sensor were used. The actual PAD was made through a wax-printing technique, and the sensor was based on an origami concept made of two layers, as can be seen in Fig. 2.1. The base consisted of four wax-defined channels extending outward from the sample reservoir (6 mm i.d.) and the top layer which had four detection and control zones (4 mm i.d.). The sample reservoir at the top was fully punched to the bottom layer, and the top was folded over. Together with a polydimethylsiloxane (PDMS) lid and a 6 mm diameter hole over the reservoir, they were held together. Eight 4 mm holes and control zones were aligned to maintain a constant pressure across the surface where the acpcPNA probe and AgNPs solution were included. The sample solution was added to the sample reservoir where it flows through the channels to wet the colorimetric detection zones. These zones were obtained by placing 10 μL of AgNPs in 0.1 M phosphate buffer saline (PBS)

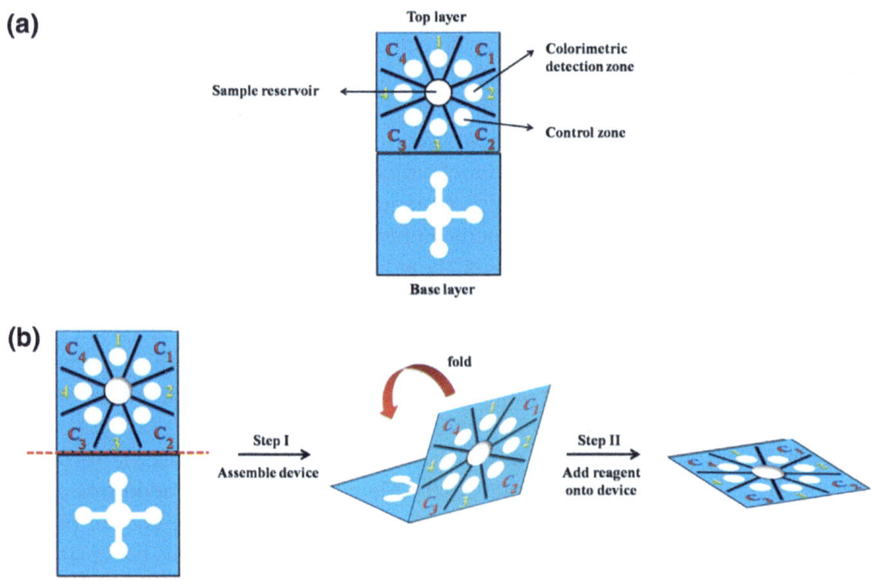

Fig. 2.1 Design and setup of paper-based multiplex DNA sensor (Teengam et al. 2017)

pH 7.4 with a ratio of 5:1 (AgNPs:PBS). For the colorimetry, special acpcPNA probes were designed and fabricated to detect synthetic oligonucleotide targets with sequences in MERS-CoV, MTB, and HPV (DNA_{com}). The intensity of the color was compared to a single-base mismatch (DNA_{m1}), two-base mismatch (DNA_{m2}), and DNA_{nc} sequences. In the presence of the DNA_{com} the intensity decreased and was unaffected by the mismatched and noncomplementary targets. There was a high selectivity to single-base mismatch, two-base mismatch, and noncomplementary target DNA.

Paper-based enzyme-linked immunosorbent assays (P-ELISAs) were created by Murdock et al. (2013) in order to measure biomolecules concentrations. In comparison to regular ELISAs, P-ELISAs are faster to make and obtain results. Neuropeptide Y was the point of interest as it is related to regulating stress, anxiety, fear, and overall sympathetic nervous system activity (Eaton et al. 2007; Heilig 2004). The target of this study was diagnosing Post Traumatic Syndrome and identifying the difference between more exposed soldiers from beginners. The platform of the P-ELISA consisted of a wax-printed 96 (12 by 8 arrays of circular test zones)-microzone paper plate, designed on Office PowerPoint according to the standard Costar 96-well microtiter plate (Murdock et al. 2013). The wax covered the areas between wells, leaving the 5.56 mm diameter wells hollowed. 3 μL of target solution was inserted into each well, followed by blocking and addition of antibody and incubation. Each test zone was 3 mm in diameter and needed 1.5 μL to be damp. Gold nanoparticles (AuNPs) of 16 nm were added to poly(ethylene glycol) (PEG) and cleaned by centrifugation and buffer exchanges. Anti-rabbit IgG antibodies were linked to the carboxylic end of the PEG. For the P-ELISA, the same process was followed except for the AuNP-IgG which was included instead of the conventional antibody through a silver enhancement kit (Ted Pella/BBInternational). The operation was tested by means of a standard 96-well plate-based ELISA procedure detecting rabbit IgG with a colorimetric substrate (Fig. 2.2). The reason behind using wax-printed paper-based was to create an inexpensive method for carrying out biomolecular assays in small volumes. The device allowed the limit of detection (LOD) to be reduced from nano to picomolar scale. Additionally, it permitted a broader range of colorimetric substances be used since the dynamic imaging ranges through conversion to grayscale. Hence, a portable device camera can be used instead of laboratory equipment to carry out the read out. The device increased the number of samples analyzed per dollar unit typically spent on diagnosis while equally increased the number of patients helped. This device holds great promises for its application in remote or resource-limited areas.

Choi et al. (2016) developed an integrated paper-based sample-to-answer biosensor for nucleic acid extraction and amplification at the POC. This provided a new view to the operation of paper-based devices as the readout could be done through visual detection or quantification using a smartphone. The device was credited as a high performance microdevice since the colorimetric detection by the naked eye could be performed within an hour. Following this strategy, a battery-powered heating device was introduced to amplify the nucleic acid in POC, which, coupled with the assay, offered a rapid target detection. A Fast Technology Analysis (FTA)

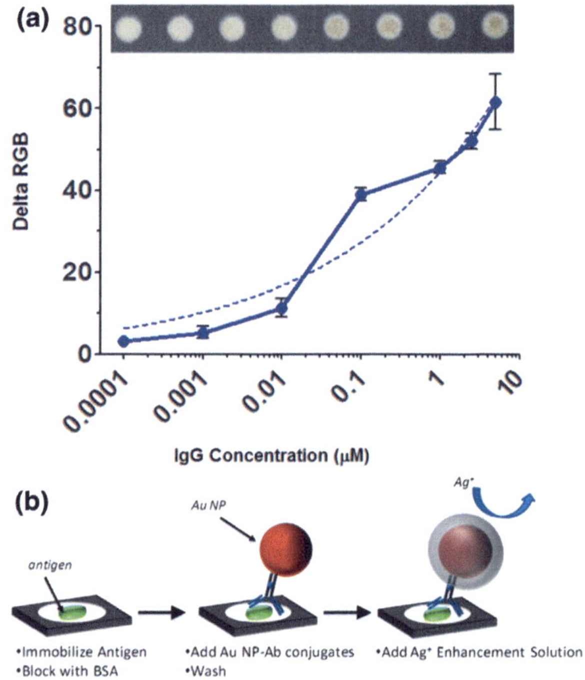

Fig. 2.2 **a** RGB versus IgG concentration using Au NP-silver enhancement procedure. **b** Enzyme-free P-ELISA assay (Murdock et al. 2013)

card and glass fiber were added to a lateral flow strip for nucleic acid extraction and amplification. First, the paper matrices were separated by hydrophobic polyvinyl chloride (PVC) layers, or valves, shown in Fig. 2.3. These valves controlled the flow from the nucleic acid extraction to the amplification zone and lateral flow strip. The device had three microfluidic layers in total: first layer for the injection holes and microfluidic channels for reagent transportation, second held the micropatterns for DNA extraction and amplification, and third incorporated a lateral flow strip for the colorimetric detection. The micropatterns were obtained from polycarbonate (PC) sheets and the microstructures from a poly(methyl methacrylate) (PMMA) sheet using a CNC milling machine. A double-sided adhesive film was used to assemble these layers. Lastly, the lateral flow strips were placed between the last two layers. The heating device was included to the integrated biosensor for very sensitive and specific loop-mediated isothermal amplification (LAMP). The bacteria were lysed outside of the device before being introduced to the inlet. This was filled in a channel by capillary forces and the debris in the microbead-bed channel flashed out to the waste chamber. Subsequently, the washing buffer erupt, forcing the solution to go through the microbeads and carry the adsorbed DNA and reaction mix into the LAMP chamber. The reaction is performed at 66 °C for 50 min and later loaded onto the lateral flow strip through the connecting channels. The developed biosensor in this study detected Escherichia coli (E. coli) in several food types with LOD as low as 10

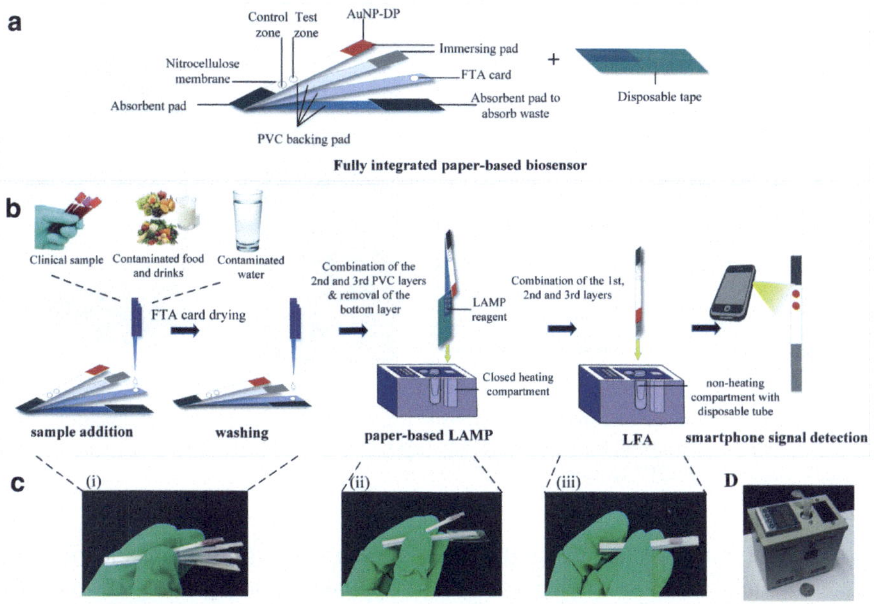

Fig. 2.3 **a** Integrated paper-based biosensor, **b** experimental procedure, **c** Biosensor during (i) extraction, (ii) amplification and (iii) lateral flow detection, **d** Handheld heating device (Choi 2016)

to 1000 CFU mL^{-1} and Streptococcus pneumonia in blood samples. Hence, proving the potentials in medical, food safety and environmental applications.

Due to the worldwide use of smartphones, Oncescu et al. (2013) created a health accessory for colorimetric detection of biomarkers in sweat and saliva based on the fact that the pH in saliva can be used to point out enamel decalcification and the pH in sweat helps indicate dehydration. The device is a noninvasive real-time analysis with disposable test strips that is connected to the phone. The pH sensing system consisted of a smartphone case, application, and test strips. The case has a slot where the strips could be inserted to be analyzed and was 3D printed from opaque Vera black material to isolate the strip form variable external light. The colorimetric analysis took place with the help of the phone's camera and a storage compartment for up to six strips. The strips were 3D printed to include an indicator strip, reference strip, and a flash diffuser. The first strip was 9 × 4 mm and was cut out from a pHydrion Spectral 5.0–9.0 plastic pH indicator strip for sweat testing and a 1.0–14.0 strip for saliva. The second strip was made of white plastic material and its purpose was to detect changes in white balance on the camera by the different light conditions or user error. The latter strip was a 2 mm thick membrane of PDMS used to minimize variations in the reading for different lighting conditions, allowing light from the camera's flash to diffuse and illuminate the posterior part of the test strip equally. Additional to the hardware, a software app was made for image acquisition and processing, and data storage and manipulation. The system as a whole, shown in Fig. 2.4, worked by first loading the app and selecting the test strip of different biomarker tests. Once the app

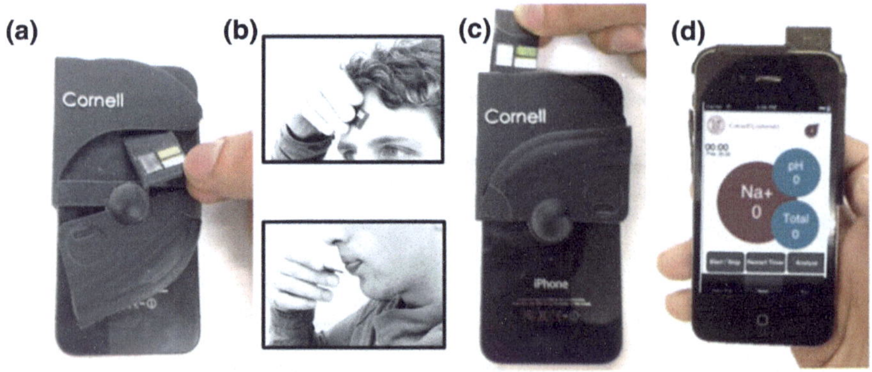

Fig. 2.4 **a** Device with the test strip being removed from storage compartment, **b** Obtaining sweat and saliva samples, **c** inserting test strip into optical system for reading, **d** pH analysis with test strip inserted Oncescu et al. (2013)

loads the calibration data and user interface, the test strip was to be inserted into the case and touch "Analyze" on the screen. The app takes a flashed image of the strip and then is categorized by color.

Matějovský and Pitschmann have created an addition from glass nanofibers to the Detehit biosensor (Pitschmann et al. 2018) for cholinesterase inhibitors (Matějovský and Pitschmann 2018). These inhibitors interfere with the nerve impulses cholinergic transfer mechanism. The biosensor was made based on the cholinesterase reaction based on enzymatic degradation of the substrate to obtain the appropriate acid or thiocholine. The Detehit biosensor was a detection ribbon based on hardened PVC and contains a detection fabric with immobilized and stabilized acetylcholinesterase (AChE), and a cellulose paper strip with acetylthiocholine iodide (ATChI) and Ellman's reagent, shown in Fig. 2.5. For its operation, the detection fabric ought to be moist and exposed to air, by placing the fabric in contaminated water or pressing a wet detection zone against the test surface. The cellulose strip was squeezed with the exposed detection fabric, making the color change visible to be analyzed. If the white detection fabric changed to yellow, there were no inhibitors, however, if the color did not change, there were inhibitors present. This is depicted in Fig. 2.6a. However, because the color differentiation was sometimes difficult, a new substrate carrier made of glass nanofibers and using a chromogenic reagent (Ellman's reagent) was developed to increase the intensity of the yellow color. The two were compared by having Ellman's reagent combined with the ATChI and an alternate butyrylthiocholine iodide (BuTChI) and testing the new device with AChE and butyrylcholinesterase (BuChE). It was a 10 cm long by 1 cm wide plastic strip, having an indication fabric on one end with an immobilized enzyme covering about 1 cm^2, while the carrier was at the other end impregnated with a substrate and an indicator, also measuring 1 cm^2, as seen in Fig. 2.6b. The detection fabric was made by impregnating a white cellulose fabric with a solution containing the enzyme (AChE tissue, AChE, or BuChE) with total activity of 21 nkat/mL, 5% of dextran, and 2%

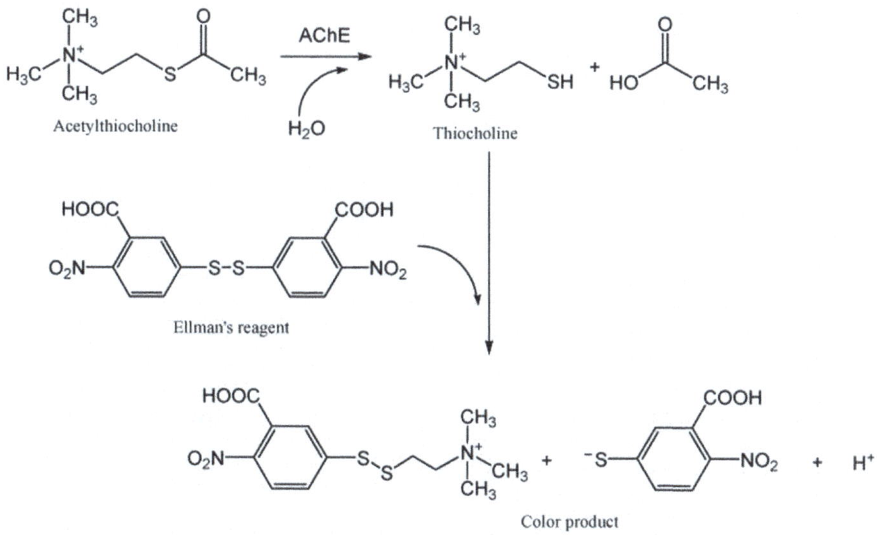

Fig. 2.5 Schematic of reaction in the Detehit biosensor (Matějovský and Pitschmann 2018)

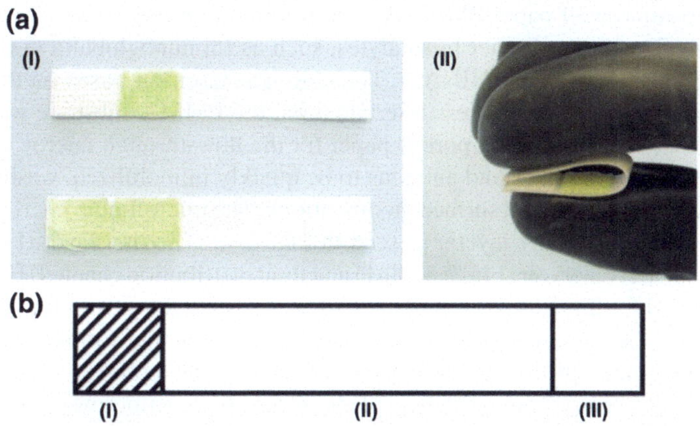

Fig. 2.6 **a**(I) Detehit biosensor before and after the test with a negative result, **a**(II) connection of opposite zones after incubation of Detehit biosensor. **b** Diagram of the biosensor, **b**(I) detection fabric; **b**(II) plastic strip; **b**(III) carrier of substrate and indic (Matějovský and Pitschmann 2018)

of anionic tenside, in a phosphate buffer solution with a 7.6 pH to be later dried at 25 °C for 24 h. The glass and cellulose papers were impregnated with a 4.3 mmol/L solution of Ellman's reagent and with 6 mmol/L of ATChI or BuTChI in ethanol. The indicator paper was dried for 6 h at 25 °C. A blank test was performed to compare, and it showed that the glass nanofibers provided an augmented color effect, as is shown in Fig. 2.7.

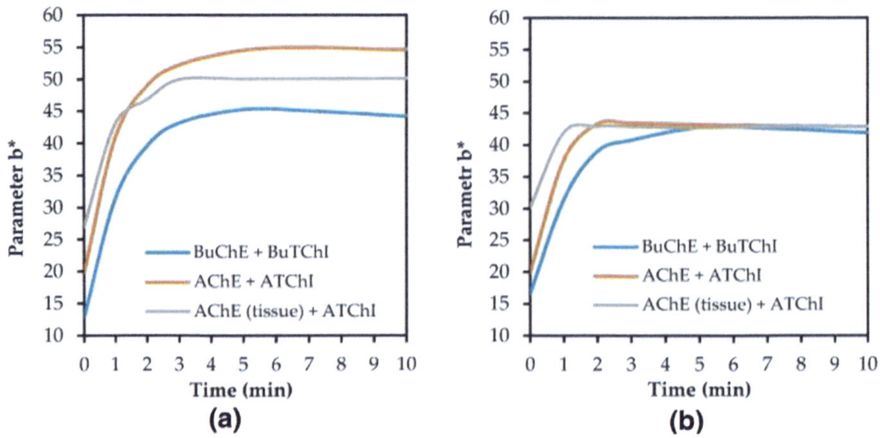

Fig. 2.7 Development of color change in the detection zone in (**a**) filter glass paper, **b** filter cellulose paper (Matějovský and Pitschmann 2018)

Recent examples of micro paper-based devices have been composed of 96 microfluidic wells (Sun et al. 2010; Sapsford 2009; Kai 2012). Sanjay et al. (2016) created a 56-microwell paper/PMMA hybrid microfluidic microplate for detection of infectious diseases and other bioanalytes, such as Immunoglobulin G (IgG) and Hepatitis B surface Antigen (HBsAg). The chip was laser cut based on the Adobe Illustrator design. In the mask-less laser ablation, the PMMA substrate was placed on a stage. The choice of using porous paper for the flow-through microwells in the PAD allowed the antibodies and antigens to be quickly immobilized, washed effectively, and avoid complicated surface modifications. The microfluidic microplate was composed of three PMMA layers, as seen in Fig. 2.8a. The first layer (Fig. 2.8bI) consists of an inlet reservoir (Fig. 2.82.8b1) and fluid distribution channel (Fig. 2.8b3) which was used for fluid delivery. It delivered the assay reagents to multiple microwells, which avoids manual pipetting and costly machinery, and forms the cover for the microwells in the following layer. Every channel in the top layer was connected to different inlet reservoirs and delivered the reagents to 7 microwells to the next layer. The second layer (Fig. 2.8bII) was for incubation and made up of 56 2 $\times$ 0.3 mm funnel-shaped microwells (Fig. 2.8c), with an upper microwell (Fig. 2.8b4) and lower microwell (Fig. 2.8b6). Paper disks (Fig. 2.8b5) were placed in between the two parts of the microwells. The microwells were created within a few minutes with a simple laser ablation method. This method offers a quick prototyping for developing microfluidic devices by means of high intensity laser beams that evaporate polymers at the focal point. Varying the intensity results in microstructures with different depths. the paper was held in place and prevented backflow of reagents as it is where the antigen or antibody were immobilized. The bottom layer (Fig. 2.8bIII) was fluid removal by means of the outlet channel (Fig. 2.8b7) leading to a common outlet reservoir (Fig. 2.8b8). Each channel was connected to a single outlet microwell to act as an outlet reservoir with a negative pressure. For the color change in HBsAg,

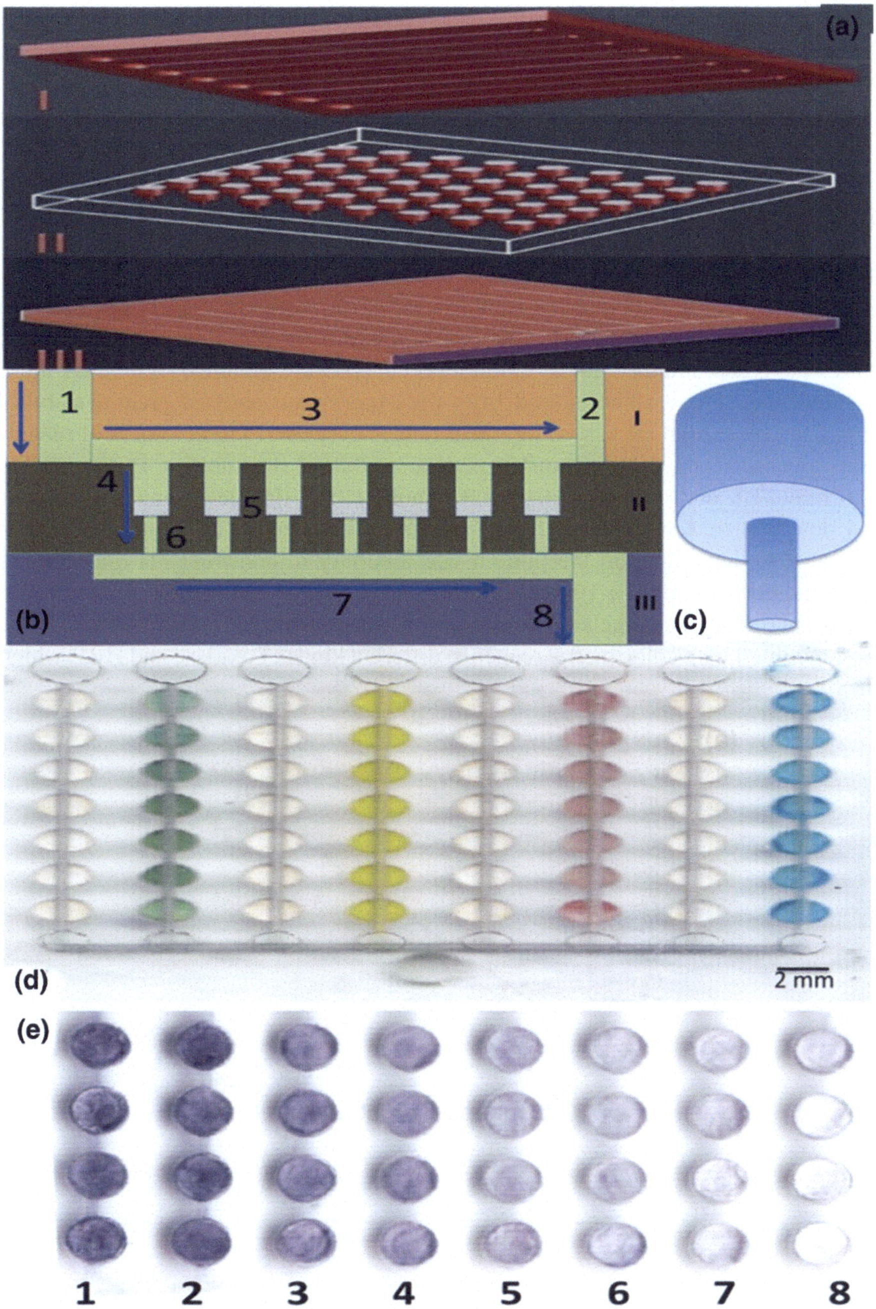

Fig. 2.8 **a** Schematic of the hybrid device, **b** cross-section of the device, **c** 3DFunnel-shaped microwell, **d** assembled device with different dyed water, **e** colorimetric representation of different concentrations of IgG (Sanjay et al. 2016)

the antigen was immobilized on the paper surface of the microfluidic microplate, reacting with the primary antibody conjugated with alkaline phosphatase (ALP). The enzymatic reaction between ALP and the colorimetric substrate BCIP/NBT was what produces the purple color. The colorimetric result could be observed by the naked eye within an hour or could be alternatively scanned by an office scanner for quantitative analysis. Figure 2.8e shows the variety of purple shades that corresponded to the concentrations inserted; the highest IgG concentrations resulted in a darker shade and as that quantity of IgG was decreased, so did the color intensity as is shown from left to right in the image.

Li et al. (2018) developed a double-layered microfluidic paper-based device with multiple colorimetric indicators for simultaneous detection of glucose, uric acid, lactate and choline. Linear calibration curves were obtained to identify these biomolecules. These values found from the experiments showed great sensibility (Fig. 2.9a) by exhibiting very wide linear ranges over two to three orders of magnitude: glucose (0.01–10.0 mmol/L), uric acid (0.01–5.0 mmol/L), lactate (0.04–10.0 mmol/L), and choline (0.04–24.0 mmol/L). The double-layered μPAD was first designed in AutoCAD. Different patterns were needed as the top layer was dedicated to detection and the bottom was auxiliary to construct 3D microfluidic channels. For detection, a 10 mm central sampling zone surrounded by eight 3 × 8 mm microfluidic channels and eight 6 mm detection zones were created. These were modified with colorimetric reagents, different kinds of oxidase and HRP, which

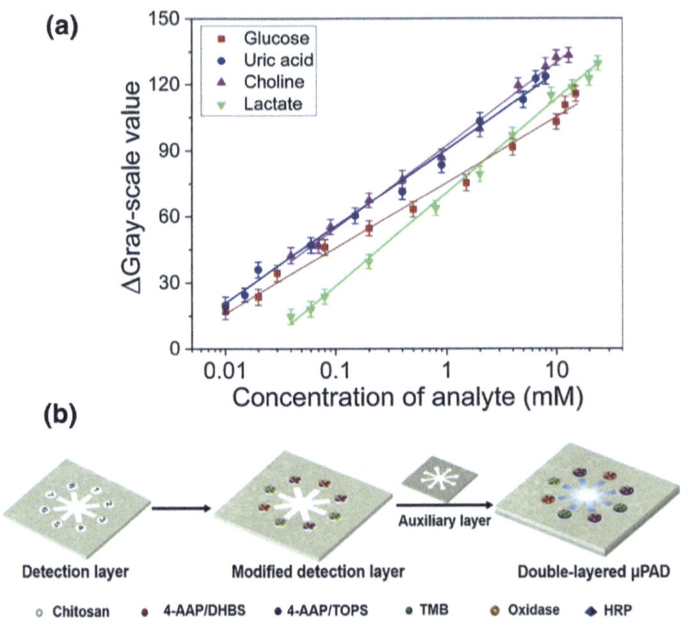

Fig. 2.9 **a** Calibration curves of glucose, uric acid, choline and lactate. **b** Fabrication of double layer μPAD Li et al. (2018)

can be observed in Fig. 2.9b. The immobilized chromogenic reagents, once oxidized by the H_2O_2 from enzymatic reactions between the oxidases and the corresponding substrates, resulted in the color change with co-immobilized HRP as catalyst. The auxiliary layer was made of one 10 mm central sampling zone and eight 3 × 10 mm microfluidic channels connected with eight 6 mm sampling zones. Also, it provided a solution connection by 3D microfluidic channels resulting from overlapping the microfluidic channels and detection zones from the top layer. A traditional wax-screen-printing technique was used to produce the hydrophilic microchannels and hydrophobic barrier on the detection and auxiliary layers. In order to prove the use, a blood sample was introduced into the sampling zone. It was then passed through the hydrophobic channels in order to react with the reactants, thus producing the color, which the Image J software could read. As can also be seen in Fig. 2.9b, two kinds of colorimetric indicators were used for each biomolecule in order to widen the detection range. This new bilayer microfluidic PAD proved to have a strong colorimetric performance, enhanced sensitivity and extended detection range.

Nilghaz et al. (2019) incorporated metal complexation to a μPAD in order to identify antibiotic residues such as oxytetracycline and norfloxacin in pork. This was done by employing the filtration quality of paper combined with aggregation and precipitation of chemical reagents. Ultimately, these processes allowed a LOD and easy result interpretation. For antibiotic residue detection, three layers of filter paper were inserted into a hydrophobic wax paper holder. The topmost layer was made from chromatography paper to serve as the detection zone. In order to detect antibiotic residues, a base substrate made from Whatman #1 and #4 chromatography paper with printed letter channels of both substances from hydrophobic wax paper, was functionalized with copper sulfate pentahydrate in 0.5 M sodium hydroxide and iron nitrate nanohydrate (colorant reagent for oxytetracycline) in a 5 mM ammonia solution (colorant reagent for norfloxacin). A transition metal hydroxide formed when a reaction occurred, allowing the residues to bind to the metal ions through coordination chemistry. In Fig. 2.10a, a schematic of the individual devices for each antibiotic residue detection can be observed. This complex coupling could result on the filter paper and provided a visible color change as the concentration increased: oxytetracycline was detected with a blue to green color change, while norfloxacin with brown to orange, as can be seen in Fig. 2.10b. The other two layers were of Whatman #4 filter paper as they were absorbance layers, meant to remove the residual liquid under the base substrate. It is important to note that the colorimetric reagents from the first layer could not diffuse into the bottom layer. The LOD for either was 1 ppm and the recovery rate for oxytetracycline was approximately 88.6% while for norfloxacin recorded to be 111.3%. The whole process of assembly and testing required less than an hour, resulting in a sensitive and rapid method to detect antibiotic residues in food samples. Since the reactions were not interfered by other antibiotics, this device can be implemented to detect other antibiotics from the same families including tetracycline and floxacin. Furthermore, The device has proven to be valuable to food safety surveillance and suitable for large-scale production.

As a common biomolecule for detection, glucose was measured from tear samples in the μPAD biosensor Moreira et al. (Gabriel et al. 2017) designed. The chromogenic

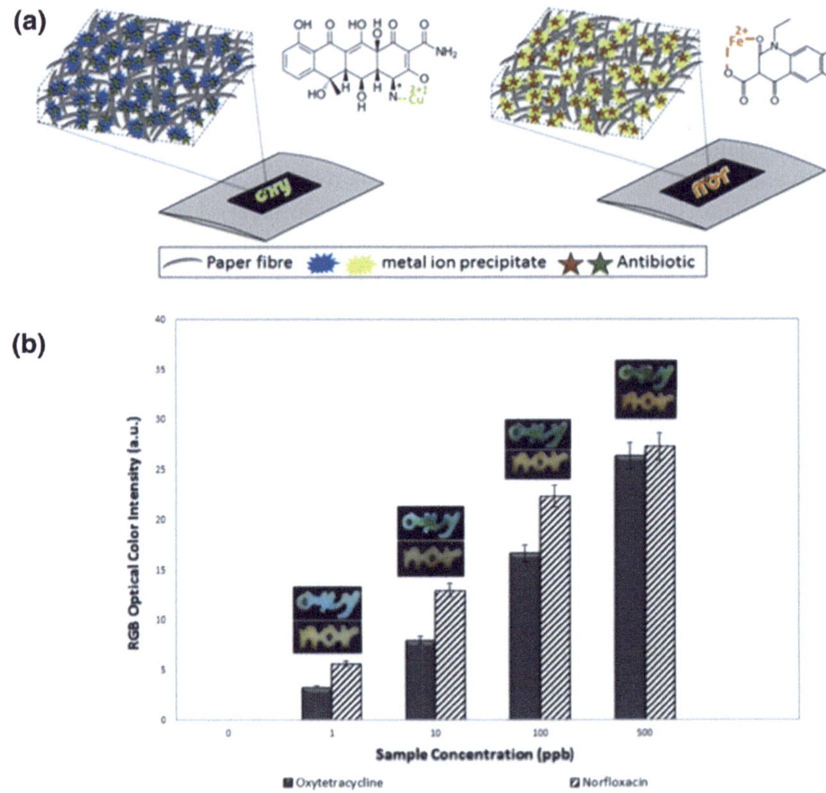

Fig. 2.10 **a** Chemical modifications of the μPAD for performing oxytetracycline and norfloxacin assays. **b** Colorimetric determination according to concentration of oxytetracycline and norfloxacin (Nilghaz and Lu 2019)

reagent used for the samples was 3,3′,5,5′-tetramethylbenzidine (TMB). The device resulted in a linear behavior between 0.1 and 1.0 mM, as seen in Fig. 2.11a, analytical sensitivity of 84 AU/mM and LOD of 50 μM. This provides an alternative for diabetic patients pricking their fingers with a lower potential interference, non-invasive, and pain-free sample (Cha et al. 2014). In order to detect glucose from tears, the desired geometry of the device was designed on Corel DrawTM graphical software and, like other μPADs, was printed on paper substrates by a wax printer (Gabriel et al. 2017). Effective hydrophobic barriers were fabricated by melting the printer wax while one side of the device was covered with adhesive tape to prevent leaking of the samples. With the basic structure, two 5 mm circular zones, identified as the control and detection zones, and a square region as the sample inlet were defined. The control zone's purpose was to detect potential interferant compounds and to minimize the matrix effect. The sample inlet is evidently where the tears are places to be pulled up to the detection zone by capillary forces. All three zones are connected by a 14 × 2 mm microfluidic channel, whereas the entire device is 24 × 10 mm. The paper

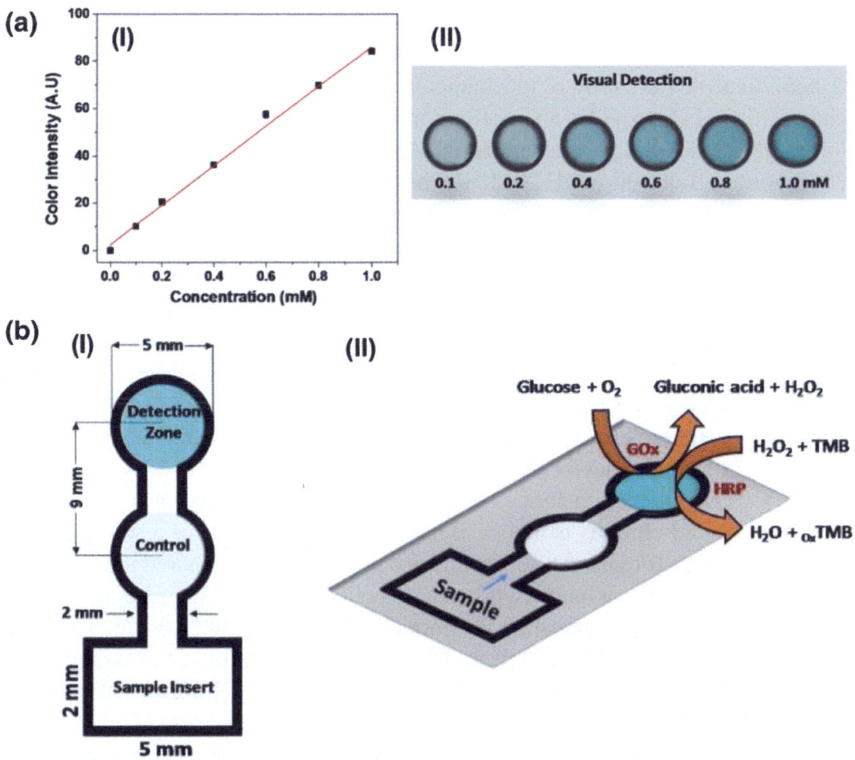

Fig. 2.11 **a** Variation of color according to the concentration of glucose in tear sample. **b**(I) Layout of μPAD for glucose colorimetric assays and **b**(II) Enzymatic reaction in presence of chromogenic reagent (TMB) for detection of glucose (Gabriel et al. 2017)

surface was modified with chitosan in order to enhance the surface attachment of enzymes. The chitosan was first prepared in 2% (v/v) acid acetic, subsequently 2 μL of the solution was introduced to the control and detection zones and allowed to dry. The detection zone was spotted with a chromogenic solution of 15 mM of TMB and 120 U mL^{-1} of an enzymatic mixture of GOx and 30 U mL^{-1} HRP. The control zone was only spotted with the enzymatic solution. 5 μL of sample aliquots were introduced to the sample inlet and left to reach the detection zone under lateral flow. This can be visualized in Fig. 2.11b. The actual colorimetric detection was done with an office HP scanner with a 600-dpi resolution. Images were taken 15 min after sample addition and were converted to Red–Green–Blue scale for simpler analysis within the Corel Photo-PaintTM software. The color intensity was directly proportional to the concentration of glucose, however, most importantly, it was compared to a personal glucometer, and no statistical difference existed with a confidence level of 95%.

Among different μPADs, Nilghaz et al. (2015) created a compact embeddable microfluidic cloth-based analytical device (μCAD) in order to detect glucose, nitrite and proteins with the naked eye and with concentrations as low as 0.5 mM, 30 μM, and 0.8 mg/dL, respectively. The device proved to be mechanically durable, robust, and flexible (Parikesit 2012). Cotton was chosen as the raw material for the cloth-based analytical device as it is mechanically robust, deliverable to the end user (Nilghaz et al. 2011), provides an excellent immobilization matrix for biomolecules (Malon et al. 2014), and a better uniform mixing of reagents and analyte through detection zones (Ballerini et al. 2011; Reches et al. 2010). Additionally, it can be easily patterned with adhesive wax to create the hydrophobic-wall microfluidic channels. Both wax and cloth are inexpensive Bhandari et al. (2011) and environmentally friendly structural material for disposable diagnostic assays (Park et al. 2004). Also, cloth-based microfluidic channels can be stable for one week at ambient temperature, making it an optimal factor for application and use in underdeveloped areas (Nilghaz et al. 2015). Overall, the instrument is a one wax-patterned cloth layer double-inlet device that includes 11 sections among the inlet points, stock zones, detection zones and isolator layers (Nilghaz et al. 2015). In order to create the 3D colorimetric microfluidic device, the 2D pattern was folded along certain predefined lines. The stock and detection zones were placed in the middle layers and separated by wax-impregnated cloth as isolators. Between 0.1 and 0.5 μL of a solution with colorimetric reagents for glucose, nitrite and protein assays were poured into multiple detection zones by a micropipette, while the detection zone held the reagents for the assay. The traditional wax patterning technique was used to pattern the microfluidic channels on scoured cotton cloth fabric. Furthermore, the ability of wax-patterned cloth fabric with hydrophilic or hydrophobic sections in order to have various designs for multiple bioassays was explored. By stacking layers of individual assay within a small surface area, and separating them by wax-impregnated fabric, multiple assays were able to be conducted. Further improvement was attempted by having an on-chip colorimetric calibration by having predefined serially diluted samples next to the detection zones.

Additionally, Lam et al. (2017) developed a chemically patterned μPAD (C-μPAD) by forming hydrophobic barriers using CVD of trichlorosilane (TCS) on chromatography paper. This C-μPAD allowed the measurement of glucose, tumor necrosis factor alpha (TNFα), and heavy metal nickel for point of care diagnostics. To create the structure of the C-μPAD, the desired fluidic pattern was designed in AutoCAD and cut out onto a vinyl tape. This tape was transferred to a 4.5 × 5 cm chromatography paper. In order to silanize the chromatography paper for the hydrophobic barriers, a low-pressure chamber and heat block were required. The vaporized TCS molecules penetrated the paper to bond covalently with hydroxyl groups on cellulose fibers creating an extremely stable and highly reproducible hydrophobic barriers, shown in Fig. 2.12a. The deposition of these TCS molecules depended on pressure, CVD duration, temperature, volume of TCS, and the mobility of the molecules. By controlling these variables, the chemicals traveled through the paper and uniformly immobilized throughout the paper. The patterned paper was placed on a hotplate to remove the vinyl tape to leave the hydrophilic area while other parts remained hydrophobic. This chemically patterned chromatography paper was then evaluated with color dyes, as seen in Fig. 2.12b, c. For glucose, the LOD was 13 mg/dL, which is that of a commercial glucose sensor. The LOD of TNFα was found to be

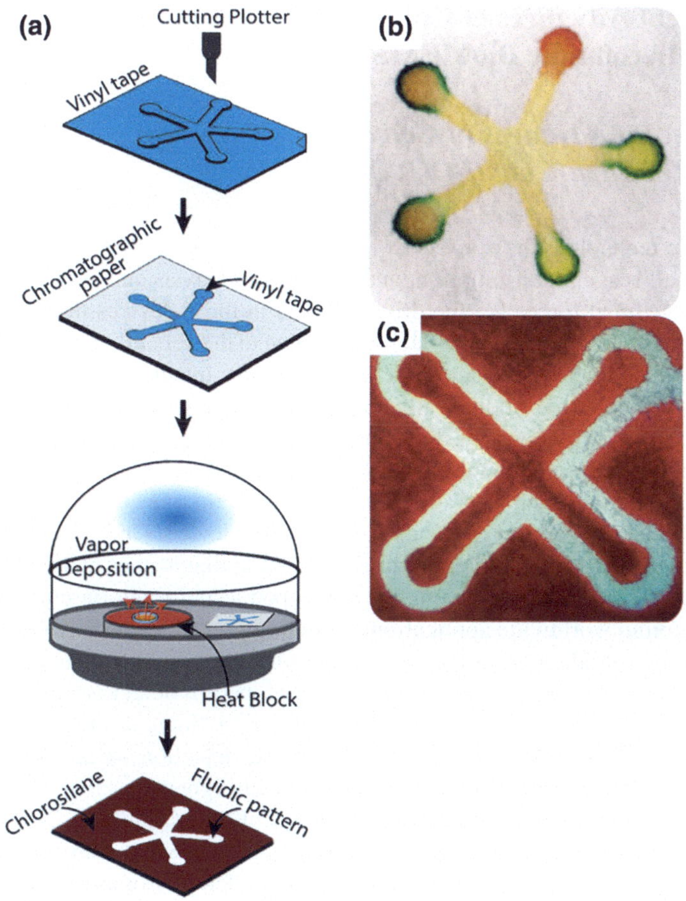

Fig. 2.12 **a** Schematic of fabrication process, **b** and **c** positive and negative features of 2D channels of C-μPAD (Lam et al. 2017)

3 ng/dL which again presented similar results as those of the commercial platforms. However, for nickel, a colorimetric agent was immobilized to obtain a stationary and uniform reaction through thermal condensation coupling method. This resulted in the detection of nickel with a LOD as low as 150 μg/dL. These LODs provided high expandability and adaptability for the device. With these results, this C-μPAD produced simple, quick, and cost-effective bioassays for environmental monitoring.

2.4 Recent Advances of Colorimetric Detection in Microfluidic BioMEMS

2.4.1 Recent Advances of Colorimetric Detection in Lab-On-Chip (LOC) Devices

Microfluidic technology has raised an increasing interest in POC diagnostics as it requires small reagent consumption, and offers fast analysis and portability. Capillary and centrifugal forces are the driving forces in these devices that have proven great candidates for integrated genetic analysis due to the versatility of fluidic control without intricate microvalves and tube lines and easy integration of the functional units (Park et al. 2016). Additionally, expensive and large laboratory set ups may be replaced by smartphones for detection analysis (Wang et al. 2016). Centrifugal microdevices usually take on the shape of a compact disc (CD) and involve a combination of microfluidic unit operations such as liquid mixing, metering, or valving which are controlled by the rotational speed of the device. Due to this versatility, many applications have forth come such as molecular diagnostics and immunoassay analysis (Sayad et al. 2017). Among the developed tests for lab-on-chip platforms, some have found worldwide applications including pregnancy tests (Li et al. 2014). Moreover, microfluidic platforms are reported for detection of Tuberculosis (Evans 2017). Furthermore, new devices are being developed to detect various pathogens for detection of foodborne diseases (Sayad et al. 2017). Some of the latest examples of the microfluidic BioMEMS used for colorimetric detection are as provided here.

Mao et al. (2017) designed a microfluidic chip with eight microchannels in order to determine chlorpyrifos based on peroxidase-like $CuFe_2O_4$/Graphene Quantum Dots magnetic nanoparticles (GQDs MNPs). The nanoparticles were included to amplify the color signal as peroxidase mimetic using a one-step hydrothermal method with electrostatic adsorption. The chlorpyrifos device was made up of a microfluidic chip with an enzyme inhibition reaction, color reaction, and UV spectrophotometric detection areas. The graphene quantum dots were synthesized from a carbonization during the pyrolysis of citric acid. A traditional soft lithography technique was used to create the microfluidic chip where a 50 μm SU-8 photoresist was spun on the silicon wafer. Subsequently, the pattern of the chip was printed on a clear film by 2880 dpi resolution ratio. The male mold of the photoresist was obtained by an ultraviolet exposure for 70 s followed by development. It was mixed with the PDMS prepolymer with a 1:10 ratio and later had air bubbles removed. The mixture was cured for 3 h under 60 °C. Later, the inlets and outlets were created on the curing PDMS substance with the microchannel structure by a puncher. Lastly, the PDMS chip was formed by plasma treatment and slide bonding. For testing, 100 μL of chlorpyrifos were injected from the first two entrances to converge at the same point and time. 100 μL of acetylcholine (ACh) and 200 μL NaH_2PO_4 buffer solutions were added to the second two entrances where the mix flew to the color area. The TMB oxidation produced the color variation and was affected by the H_2O_2 concentration with the

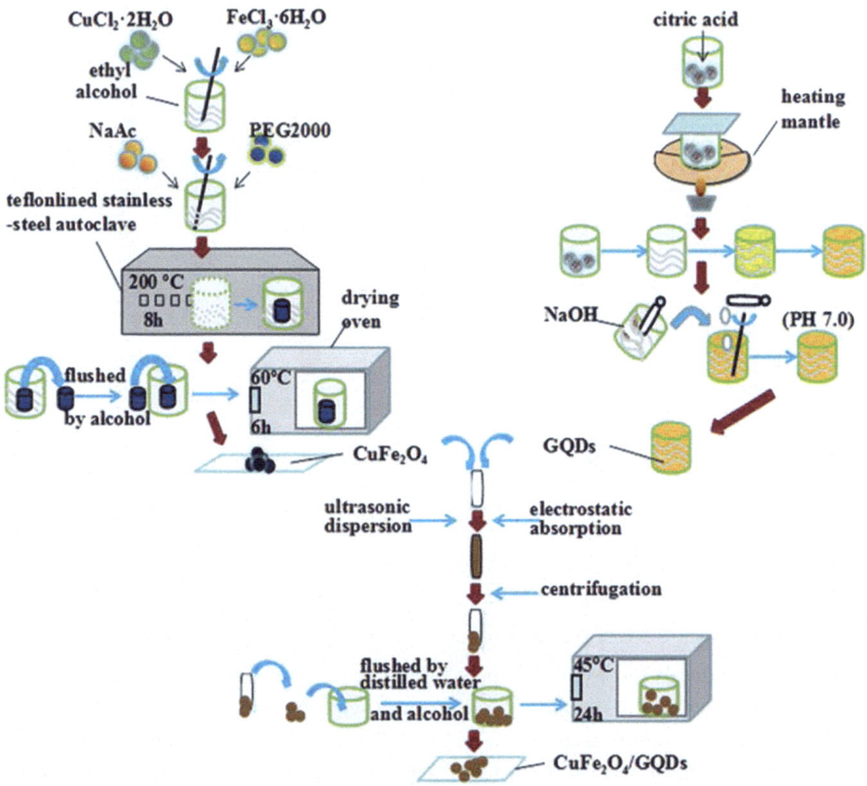

Fig. 2.13 Process of $CuFe_2O_4$/GQDs formation (Mao et al. 2017)

$CuFe_2O_4$/GQDs (Fig. 2.13). ACh was inhibited as the organophosphate pesticides (OPs) concentration increased, hence reducing the production of H_2O_2 and thus, provoking a weak color reaction and absorbance. Therefore, it was concluded that the absorbance is inversely proportional to the concentration of OPs.

Another example is a novel biosensor that uses gold nanoparticles to detect the concentration of E. coli O157:H7 equipped with an app for color monitoring (Zheng et al. 2019). The microfluidic chips have proven to detect foodborne pathogens rapidly due to its precise control of the fluids, few sampling, and decreased detection time. For the E. coli-detecting biosensor, shown in Fig. 2.14, the 3D printed, and surface plasma-bonded microfluidic chip was the most important piece. The mold of the chip was placed in 5% NaOH for half an hour and later mixed with curing agent at a ratio of 10:1. It was placed into the mold for 12 h at 65 °C and once peeled, it was united with the glass slide through surface plasmon treatment. It had two 600 × 100 μm serpentine mixing channels, where one is to mix the bacterial sample with the MNPs and polystyrene (PS) microspheres, and the other for mixing catalysate with the AuNPs and cross-linking agents. The COMSOL platform was used in these channels to stimulate them based on free triangular grid and finite volume. The chip

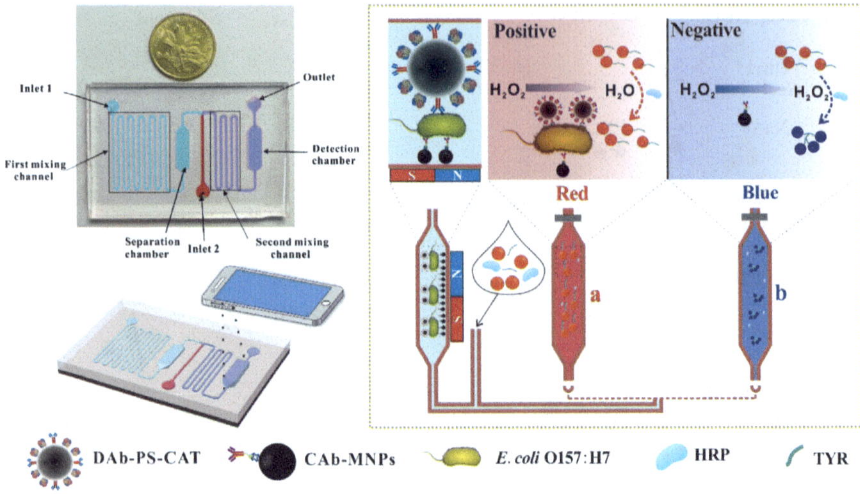

Fig. 2.14 Microfluidic colorimetric biosensor for detection of E. coli O157:H7 based on gold nanoparticles and smartphone imaging (Zheng et al. 2019)

also incorporated a 14 × 14 × 1 mm chamber that separated the MNP-bacteria-PS complexes and catalyzed hydrogen peroxide. The last part was a 14 × 14 × 2 mm detection chamber where the AuNPs color modifications were observed. The MNPs modified with the capture antibodies and the PSs modified with the detection antibodies were used to interact with the target bacteria in the first mixing channel. AuNP were added for signal indication, Hue-Saturation-Lightness (HSL)-based smartphone imaging correctly detected changes in color, and the microfluidic chip was created for on-chip bioreaction. The results showed that the LOD was 50 CFU/mL for E. coli O157:H7 and the mean recovery was -96.8%.

An analyzer was created by Li et al. for multi-index monitoring of diabetes and hyperlipidemia from a patient's blood Li et al. (2019c) The indexes for monitoring involves glucose (GLU), triglyceride (TG), and total cholesterol (TC). The color changes originated from the peroxidase-H_2O_2 enzymatic reactions and were taken with a smartphone analyzer that contained a LED light and a charge-coupled device (CCD) camera. The smartphone-assisted microfluidic analyzer contained a 2 mm thick structural layer made from plastic injection molding with pressure sensitive adhesive (PSA) layers on either side. The most important piece was the fan-shaped body of the device with 3.25 cm radius, a buffer pool, four reaction chambers with vent holes, and two positioning holes, each with three capillary stop valves (Fig. 2.15). The analyzer was an optical detection system based on step motor, microcontroller, and Bluetooth module. The detection zone included a white LED, macro lens and a CCD camera. To build the microchip, the plastic was first attached to only the bottom layer of adhesive. Subsequently, 13 μL of each detection reagent was input to each of three chambers, leaving one empty to serve as control. The reagents were left there during incubation for 4 h at 37 °C and then the top adhesive was added.

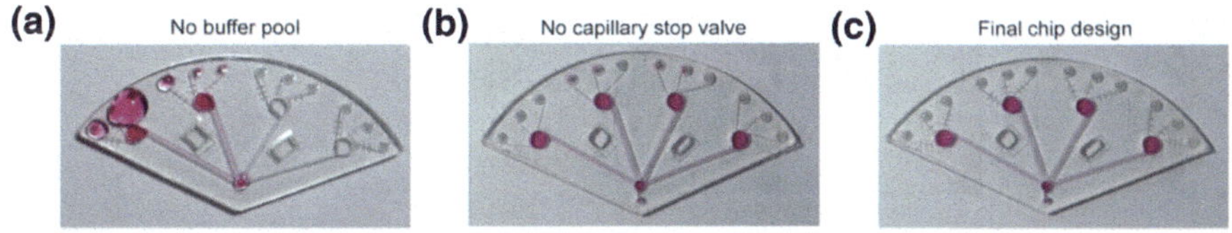

Fig. 2.15 Improvements done to the biosensor. **a** Leak from vent holes by not having buffer pools. **b** Small leak in the chip without the capillary stop valves. **c** No leakage by having the buffer pool and capillary stop valves (Li et al. 2019c)

The chip could then be sealed into a vacuum pouch and stored for 6 months at 4 °C. In order to test the system, 10 μL of serum was mixed with 190 μL of Tris–HCl and 4-Aminoantipyrine in an Eppendorf tube. 95 μL of the mixture was input to the inlet hole under the pressure of the pipette and the chip went into the Smartphone-assisted microfluidic chemistry analyzer for incubation for 15 min. The detection was done with the phone by means of the light provided by the LED and transmitted by the reagent. It was then collected by the macro lens and hence, detected by the camera as a step motor rotated the chip so all four chambers could be recorded. The images were processed by the microcontroller and sent via Bluetooth to a smartphone for the final analysis. To improve the system, the reagent addition steps could be automated to reduce the manual operation. Additionally, a cost-effective and reliable detector must be put in place to obtain such quantitative results as in Fig. 2.16. In this device, the detection reagents were mutarotase, glucose oxidase, peroxidase and 2-hydroxy-3,5-dichlorobenzenesulfonic acid (DHBS) for glucose; cholesterol esterase, cholesterol oxidase, peroxidase and DHBS for cholesterol; and lipoprotein lipase, glycerokinase, glycerol-3-phosphate oxidase, peroxidase and DHBS for triglyceride.

An Electronics-based ELISA (e-ELISA) using a Lab-on-a-Printed Circuit Board (LoPCB) device for Point of Care (POC) of Tuberculosis was developed by Evans et al. (2017). The device was a modified ELISA that operated with 10 μL volume and included PMMA wells, gold surface, TMB as the reporter reagent (Fig. 2.17a) and Interferon Gamma (IFNγ), a pro-inflammatory cytokine key in innate and acquired immunity, as the assay target. Copper (Cu) foil was laminated to the FR4 PCB substrate through thermal adhesion. Subsequently, tracks made from electrical connections, and electrode pads were patterned by etching the Cu layer. A gold layer was plated on top of the copper one and was fixed by the copper primer which determined the final distribution of gold. The fluid wells were cut from PMMA and fixed onto the surface of the circuit board. The board (Fig. 2.18a) included reference electrode circuitry, working electrodes with amplification circuitry, voltage input Analogue-to-Digital Converters (ADCs), processing unit and the user interface consisted of an embedded on-board TFT touch screen USB port (Fig. 2.18b). Both amperometric and colorimetric signals were measured by the device. The first was measured by second generation amperometry where it detected charge carrier concentration through the measurement of total current magnitude charge carriage.

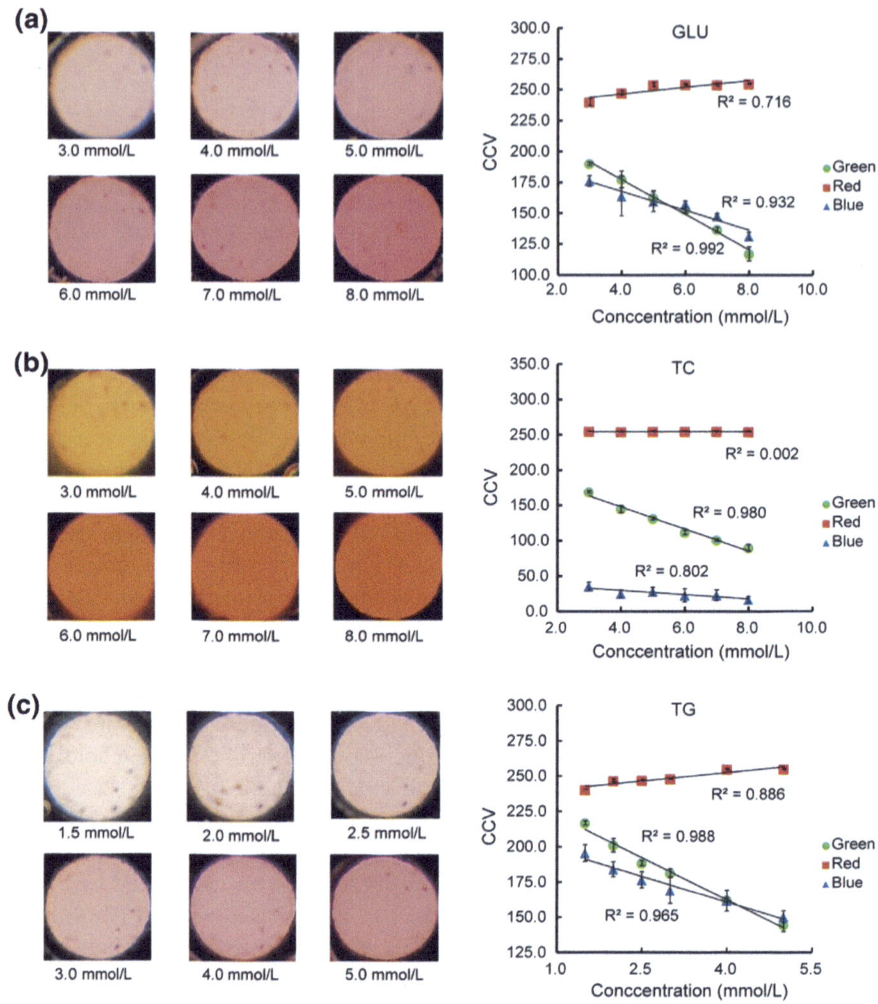

Fig. 2.16 Colorimetric results for each index (Li et al. 2019c)

To accomplish this, a reporter molecule was needed that had a relatively low conductivity. For colorimetric readout, capture antibody α IFNγ Fab′-3Cys-6His) was immobilized on the Au sensor chip surface and with the addition of cysteine residue. Covalent bonding was achieved, hence capturing and immobilizing antibody fragments by Cysteine (thiol) linkage. The color change was generated from the resulting color of TMB which was originally a colorless liquid that turned bright blue after the reaction (Fig. 2.17b). The outcome of this study was a lightweight, low-cost, amperometric and colorimetric detection unit made according to standard commercial processes which embedded microfluidics and multi-channel amperometric sensing.

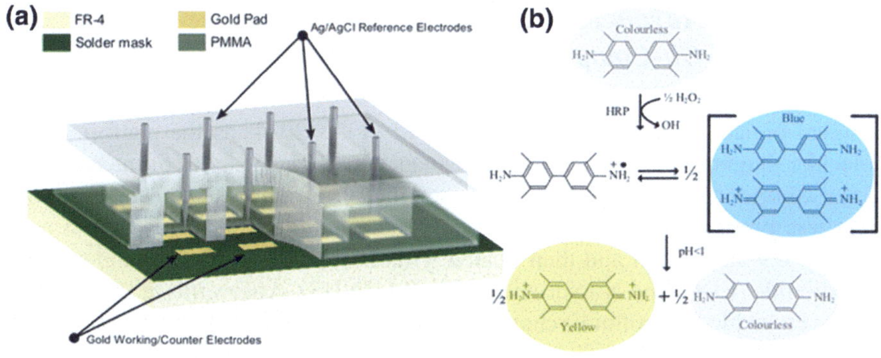

Fig. 2.17 **a** Schematic of fabricated PCB-based biosensor, **b** representation of HRP-catalyzed oxidation of TMB (Evans 2017)

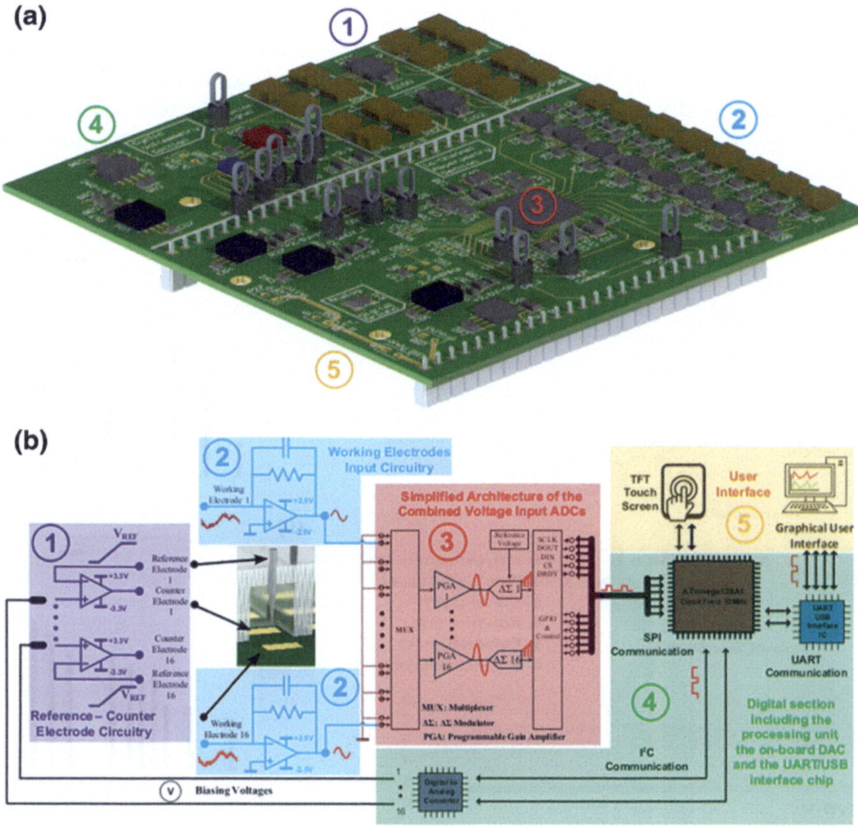

Fig. 2.18 **a** Schematic of custom-made electronic board, **b** individual sections of the electronic board and location on (**a**) (Evans 2017)

2.4.2 Recent Advances of Colorimetric Detection in Lab-On-Compact Disk (LOCD) Devices

A digital optical disc (DVD) was used by Li et al. (2014) as the platform for a molecular diagnostic and quantitative pregnancy test. The analytes of interest were human Chorionic Gonadotropin (hCG) identified from urine samples. A standard DVD was prepared for signal readout. The polycarbonate surface was first activated by UV irradiation and then treated with 1-ethyl-3-(3-dimethylaminopropyl) carbodiimide (EDC) and N-hydroxysuccinimide (NHS). Afterwards, a PDMS plate with six embedded microfluidic channels was placed inside the DVD. Once activated, anti-human Chorionic Gonadotropin Gα (anti-hCGα) monoclonal antibodies were immobilized on the surface (Fig. 2.19a) by an amide-coupling reaction. A streptavidin nanogold conjugate was added to the surface via biotin-streptavidin interaction following a silver staining treatment, which ultimately resulted in an enhanced signal in order to obtain a significant disruption of the laser readout in the optical drive. A standard, unmodified optical drive was used for the assay readout, and free disc-quality analysis software for the processing of the obtained data. To use the device, samples were initially loaded into the PDMS microfluidic channels, subsequently, the DVD was spun within the optical drive, undergoing centrifugal forces that created a different radial distance according to the analyte concentration. The assay was tested

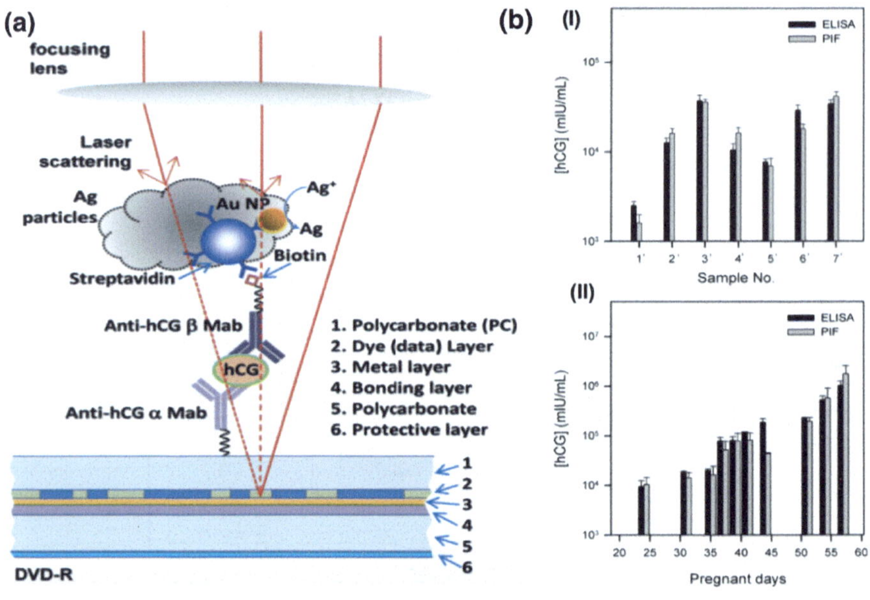

Fig. 2.19 **a** Schematic of DVD assay design and principle of signal reading. **b**(I) Quantitation of urine hCG level of seven pregnant women during different points of pregnancy. **b**(II) Quantitation of urine hCG levels of one woman during different days of pregnancy by ELISA and parity inner fails (PIF) methods (Li et al. 2014)

with a DVD diagnostic software and the readings were interpreted according to the radial distance and optical darkness ratio. The results showed comparable sensitivity and selectivity to well-established colorimetric methods and ELISA (Fig. 2.19b). Additionally, it is an inexpensive, easy to use, multiplex, POC diagnostic instrument for prompt response used in remote and/or rural areas.

Park et al. (2016) developed an integrated rotary microfluidic system (Fig. 2.20a) with DNA extraction unit [Fig. 2.20c(I)], LAMP [Fig. 2.20c(II)], and lateral flow strip [Fig. 2.20c(III)]. The device was used for detection of the food-borne bacterial pathogen, monoplex Salmonella Typhimurium and multiplex Salmonella Typhimurium as well as Vibrio parahaemolyticus. The LAMP reaction replaced

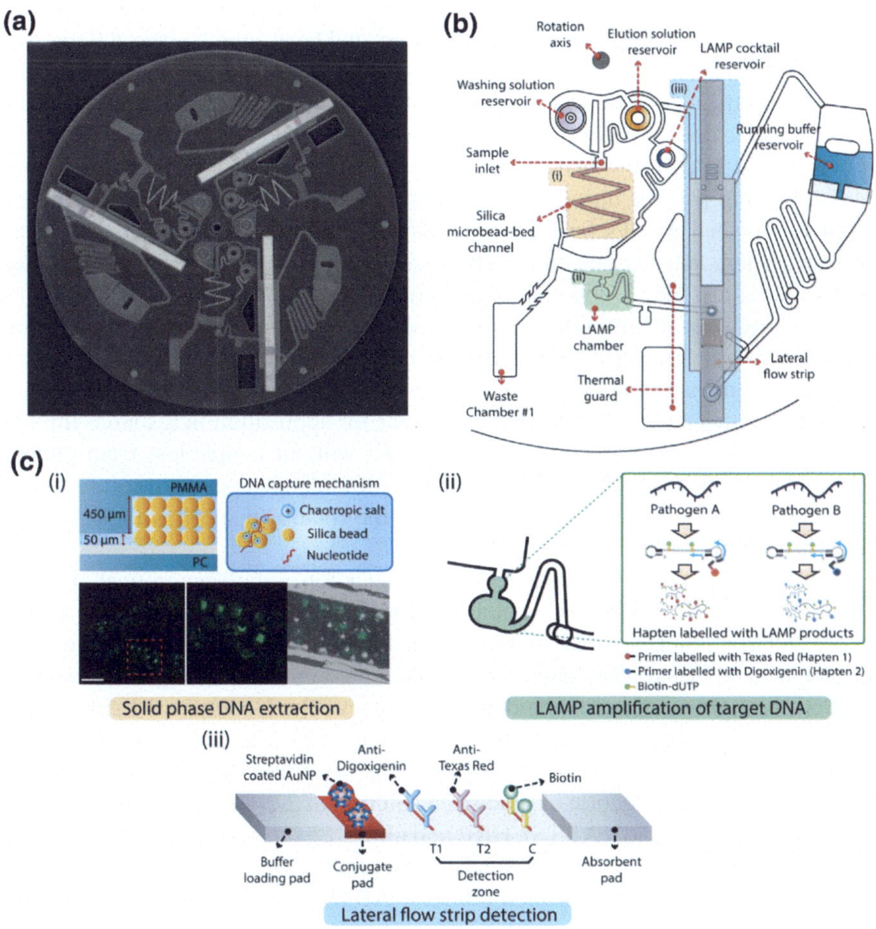

Fig. 2.20 **a** Image of integrated rotary microdevice, **b** illustration of integrated rotary microdevice, **c**(I) schematic of solid phase DNA extraction and fluorescent images of the adsorbed glass microbeads, **c**(II) schematic of LAMP amplification of target DNA, **c**(III) schematic of lateral flow strip detection (Park et al. 2016)

polymerase chain reaction (PCR), allowing higher specificity and sensitivity, and eliminating the bulky thermocycler. A glass microbead-based centrifugal nucleic acid extraction was used as a solid phase matrix where the genomic DNA could be purified from the lysate sample. Lastly, the colorimetric based lateral flow strip provided a cost-effective and equipment-free detection method. The device integrated these three techniques for detecting in a sequential manner with an optimized microfluidic design and rotational speed control. The microdevice was a five-layer stacked disc with three identical units. Each unit consisted of three functional parts: solid phase DNA extraction, LAMP reaction, and a lateral flow strip. The main three microfluidic layers were created by a CNC milling machine and PSA film was used once the micropattern was cut by a plotter, following a hot press bonding all layers together. The first layer had injection holes and microfluidic channels to transport a LAMP product and a running buffer from the second layer into the lateral flow strip on the third layer. The second layer contained the micropatterns for DNA extraction and amplification. The third layer was for embedding a lateral flow strip for the colorimetric detection (Fig. 2.20b). The detection could be made with the naked eye due to the LAMP product. In order to do so, a lateral flow strip containing a buffer loading pad for introducing a running buffer, a conjugate pad (including streptavidin coated AuNPs for the conjugation with the LAMP products), a detection zone (where anti-Digoxigenin, anti-Texas Red and biotin were immobilized in the test line 1, test line 2, and control line, respectively), and an absorbent pad for liquid wicking were incorporated within the device. Thermal guards were patterned around the lateral flow strips to prevent the heat from influencing the anti-haptens on the detection zone of the lateral flow strips. The resulting microdevice presented great potentials as a user-friendly POC analyzer for application in resource-limited setups. The design allowed efficient fluid transfer without sample loss from sample pretreatment to strip detection. The automatic and integrated genetic analysis could then be successfully performed by controlling the rotational speed without the use of expensive equipment.

A cheap and portable smartphone spectrometer for monitoring optical changes as they occur was created by Wang et al. (2016). The device was aimed at detecting glucose and troponin I, a myocardial infarction biomarker by means of a smartphone with a built-in LED and complementary metal oxide semiconductor (CMOS) camera to use as the light source and the detector, respectively. No external light source, lens, or filter were required. As the dispersive unit, a CD with grating was used. For human cardiac troponin I detection, peptide functionalized AuNPs were taken as the reporters. For the detection of glucose, a solution of 2,2′-azino-bis (3-ethylbenzthiazoline-6-sulfonic acid) (ABTS), HRP, and glucose oxidase (GOx) was utilized. A bi-enzymatic cascade assay was used where the glucose was catalytically converted into hydrogen peroxide, which then converts ABTS by HRP into oxidized form. Once oxidized, a blue color appeared and was read from the color band with the spectrometer bases on the change of intensity. The spectrometer relied on a sample cell with an integrated grating substrate, and the phone's LED flash and camera. The CD was placed 50 mm away from the LED and tilted 5° so that the flashlight passed through a 1 mm diameter pinhole. The grating tracks were aligned to the

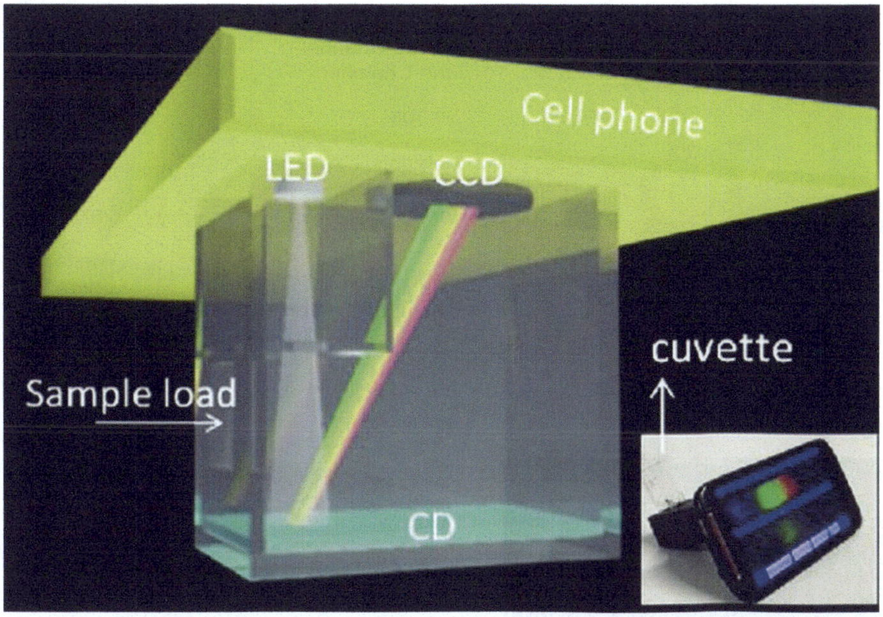

Fig. 2.21 Schematic of installation of the device (Wang 2016)

incident light, and the light was refracted from the CD onto the camera (Fig. 2.21). This allowed a real-time measurement and resulted in a LOD of 50 ng mL. The biosensing system coupled with a smartphone platform offered a promising method for the phone to detect, interpret, and communicate targeted biological information. Likewise, a higher sensitivity, speed, and simultaneous monitoring was possible. Although it had a comparable performance to commercial devices, it was more compact, cost-effective, and portable.

Sayad et al. (2017) created a 165 mm diameter centrifugal microfluidic device platform integrated with LAMP technique (Fig. 2.22a, c) for quick, monoplex and colorimetric detection of foodborne pathogens. Three main pathogens were studied: Salmonella spp, Es. coli and Vibrio cholerae. 24 strains of these pathogenic bacteria with eight strains of each bacterium were tested and DNA amplification on the microfluidic CD was performed for 60 min. The device consisted of three layers: PMMA top and bottom layers, and a PSA middle layer as shown in Fig. 2.22d. The layers were aligned and press-bounded together. The top layer contained the venting and loading holes for liquid insertion and wax plug. For the microfluidic structures in the bottom layer, six identical units (Fig. 2.22b) were designed to be able to perform 30 genetic analyses of the three pathogens. One of the units was the loading chamber which loaded the LAMP reagents and primers. Another was the mixing channel and chamber that aliquot the LAMP assay into equal volumes. The sealing chambers contained the sealing material used to seal the connection

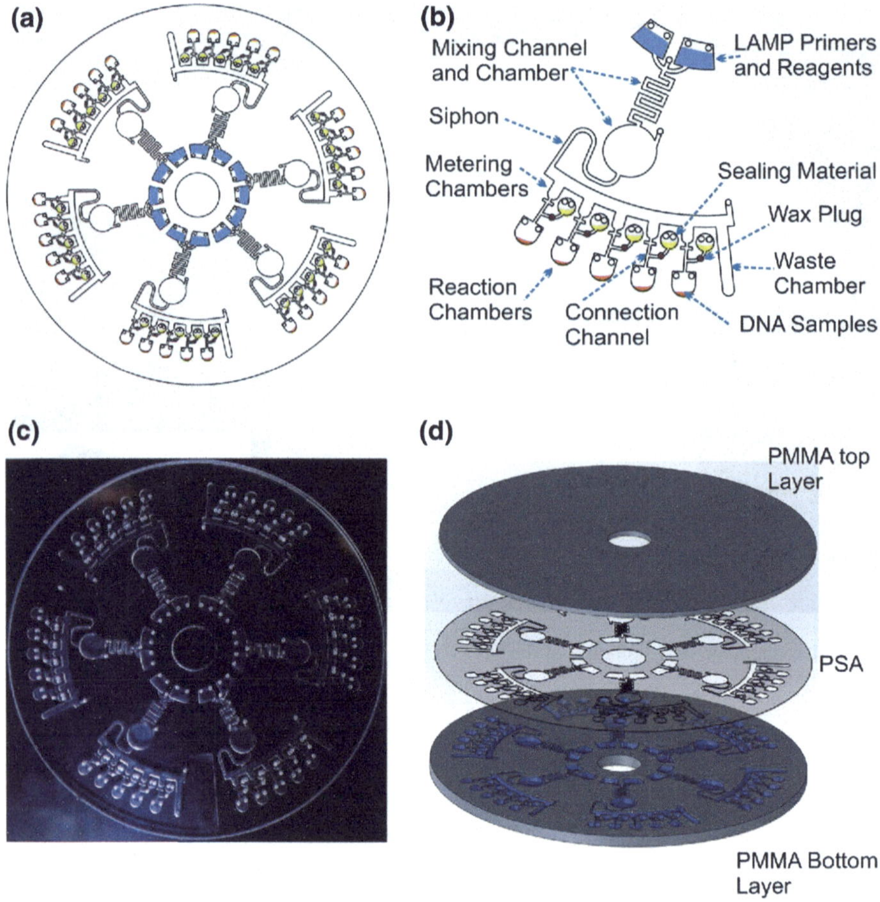

Fig. 2.22 **a** Schematic of centrifugal LAMP microdevice, **b**top view of one unit of the microdevice, **c** photograph of the microdevice, **d** schematic top view of the device's layers (Sayad et al. 2017)

channel between the metering and amplification chambers to prohibit liquid evaporation. Lastly, the amplification chamber was designed for the amplification and detection of DNA. An optimized square-wave microchannel, metering chambers, and revulsion per minute (RPM) control were utilized to constantly load, mix, and aliquot the LAMP primers/reagents as well as DNA samples. The LAMP reaction amplicons were detected by the calcein dye colorimetric method and analyzed with the developed electronic endpoint detection system (Fig. 2.23a) including the Bluetooth interface to send the results to a smartphone (Fig. 2.23b). Calcein is a synthetic fluorescein that emits a bright fluorescence creating a visual color change. A positive sample changes from yellow to green while a negative readout remains light orange. The entire process in only one CD lasted about 65 min and presented a LOD of 3×10^{-5} ng μL^{-1}.

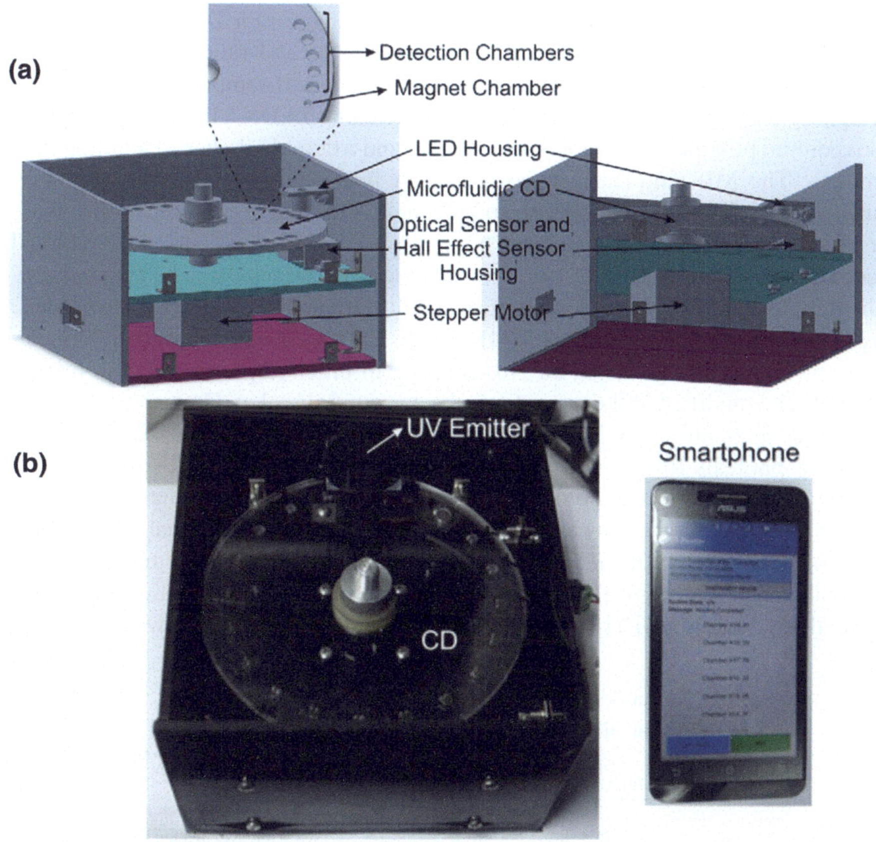

Fig. 2.23 **a** Schematic of endpoint detection system, **b** photograph of endpoint detection system and the application software on a smartphone (Sayad et al. 2017)

2.5 Alternative BioMEMS for Colorimetric Detection

In order to improve the colorimetric biosensing strategy, different amplification methods have been developed. Among these alternative strategies, exonuclease (Exo)-assisted signal amplification, strand displacement amplification (SDA), and rolling circle amplification (RCA) are promising methods. The latter has proven to result in ultrasensitive biosensors due to its excellent properties in signal amplification. However, it creates nonspecific amplification due to the impurity of the circular template, and generation of large fragments of single-stranded DNA (ssDNA) which may decrease the solubility. To counterpart this disadvantage, an improved alternative was developed (Li 2016).

Li et al. (2016) established an RCA-based colorimetric biosensor with an enhanced nucleic acid-based amplification machine to detect attomolar microRNA (miRNA). The machine was composed of a complex of trigger template and cytosine-rich

DNA co-modified molecular beacon (MB) and guanine-rich DNA (GDNA) as a probe. This was made by mixing MB and GDNA at a 1.2:1 ratio inclubated for an hour at room temperature. Seal probe was prepared by self-templated ligation of 5′-phosphorylated dumbbell-shaped DNA sequence using T4 DNA ligase. The machine also required polymerase and nicking enzyme, and a dumbbell-shaped amplification template. The MB template was composed of four sections: miRNA-recognition domain (Fig. 2.24a), GDNA hybridization domain (Fig. 2.24b), amplification domain for producing the nickel triggers (Fig. 2.24c), and a nicking domain for Nb.BbvCl recognition. The target miRNA triggered MB mediated strand displacement to cyclically release nicking triggers, leading to a toehold-initiated RCA (TIRCA) to produce large amounts of GDNAs (Fig. 2.24). These can stack with hemin to form G-quadruplex/hemin DNAzyme, an HRP mimic, in order to produce a colorimetric reaction. The modified MB decreased the background signal and improved the stringent target recognition. A DYY-6C electrophoresis analyzer was used to perform gel electrophoresis for the seal probe and a Bio-rad ChemDoc XRS for imaging. A NanoDrop 100 spectrophotometer, a UV–visible spectrophotometer, collected the signal. The outcome was a simple, label-free ultrasensitive visual colorimetric biosensor (down to a LOD of 5aM, a detection range of nine orders of magnitude for practical sample analysis). The sensitivity was due to the reduction of steric hindrance and facilitated solution of TIRCA products. The entirety of the process was completed

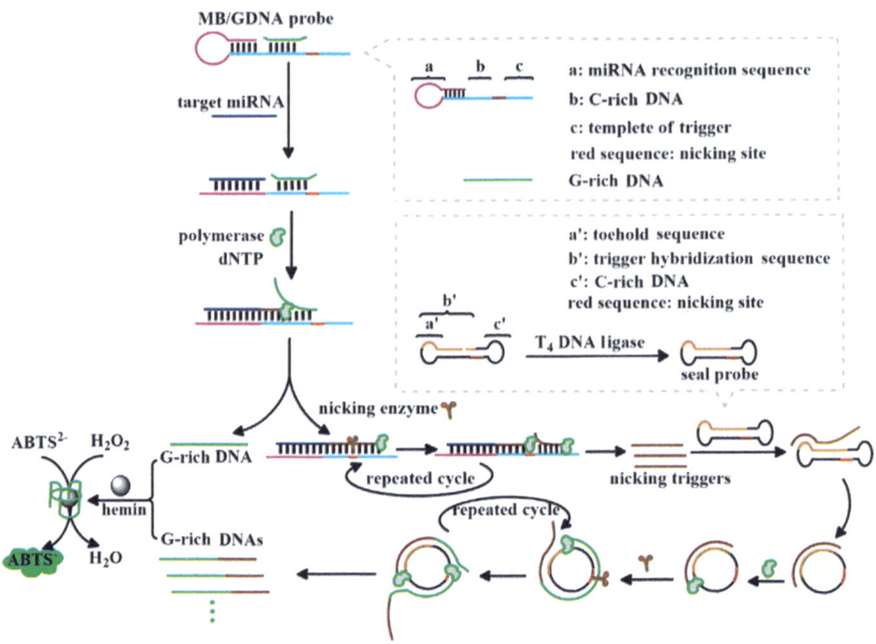

Fig. 2.24 Schematic of nucleic acid-based amplification machine (Li 2016)

within 90 min. The machine itself offered a combination of advantages provided by enzymatic signal amplification and toehold-initiated RCA.

2.6 Summary

Colorimetric detection can be done visually, hence no expensive equipment or much time is needed. Paper-based devices that operate based on colorimetric detection strategy have improved the accessibility, speed, and accuracy of tests while offering considerable cost effectiveness. Smartphones and tablets opened yet another window of opportunity to easy and onsite analysis of the readout results. By combining microfluidics with the μPADs, the advantages of these devices are as μPAD with colorimetric results attract even more attention due to its simplicity, versatility, straightforward detection results and applicability, especially in point of care analysis without advanced instruments (Li et al. 2018). Microfluidic colorimetric biosensors offer small size, high precision with small sample size, simple operation, and low cost (Mao et al. 2017). Overall, colorimetric-based enzymatic assays are fast, adaptable, and cost-effective while allowing the color change to be seen by the naked eye or by digital sensors (Li et al. 2019c). Centrifugal microfluidic devices present excellent opportunity to detect a variety of biomolecules for different applications. While the literature has witnessed a great deal of advancements in fabrication and application of BioMEMS for colorimetric biosensing, further optimization of these devices for high throughput detection present an opportunity for further improvement.

References

Aldewachi H, Chalati T, Woodroofe MN, Bricklebank N, Sharrack B, Gardiner P (2018) Gold nanoparticle-based colorimetric biosensors. Nanoscale 10(1):18–33. https://doi.org/10.1039/c7nr06367a

Ballerini DR, Li X, Shen W (2011) An inexpensive thread-based system for simple and rapid blood grouping. Anal Bioanal Chem 399(5):1869–1875. https://doi.org/10.1007/s00216-010-4588-5

Bhandari P, Narahari T, Dendukuri D (2011) Fab-chips: a versatile, fabric-based platform for low-cost, rapid and multiplexed diagnostics. Lab Chip 11(15):2493–2499. https://doi.org/10.1039/c1lc20373h

Cha KH, Jensen GC, Balijepalli AS, Cohan BE, Meyerhoff ME (2014) Evaluation of commercial glucometer test strips for potential measurement of glucose in tears. Anal Chem 86(3):1902–1908. https://doi.org/10.1021/ac4040168

Choi JR et al (2016) An integrated paper-based sample-to-answer biosensor for nucleic acid testing at the point of care. Lab Chip 16(3):611–621. https://doi.org/10.1039/c5lc01388g

Eaton K, Sallee F, Sah R (2007) Relevance of neuropeptide Y (NPY) in psychiatry. Curr Top Med Chem 7(17):1645–1659. https://doi.org/10.2174/156802607782341037

Evans D et al (2017) An assay system for point-of-care diagnosis of tuberculosis using commercially manufactured PCB technology. Sci Rep 7(1):1–10. https://doi.org/10.1038/s41598-017-00783-8

Fuwa K, Vallee BL (1963) The physical basis of analytical atomic absorption spectrometry: the pertinence of the beer-lambert law. Anal Chem 35(8):942–946. https://doi.org/10.1021/ac60201a006

Gabriel EFM, Garcia PT, Lopes FM, Coltro WKT (2017) Paper-based colorimetric biosensor for tear glucose measurements. Micromachines 8(4):1–9. https://doi.org/10.3390/mi8040104

Heilig M The NPY system in stress, anxiety and depression. Neuropeptides 38(4):213–224. https://doi.org/10.1016/J.NPEP.2004.05.002

Kai J et al (2012) A novel microfluidic microplate as the next generation assay platform for enzyme linked immunoassays (ELISA). Lab Chip 12(21):4257. https://doi.org/10.1039/c2lc40585g

Lam T, Devadhasan JP, Howse R, Kim J (2017) A chemically patterned microfluidic paper-based analytical device (C-μPAD) for point-of-care diagnostics. Sci Rep 7(1):1–10. https://doi.org/10.1038/s41598-017-01343-w

Li D et al (2016) A colorimetric biosensor for detection of attomolar microRNA with a functional nucleic acid-based amplification machine. Talanta 146:470–476. https://doi.org/10.1016/j.talanta.2015.09.010

Li X, Weng S, Ge B, Yao Z, Yu HZ (2014) DVD technology-based molecular diagnosis platform: quantitative pregnancy test on a disc. Lab Chip 14(10):1686–1694. https://doi.org/10.1039/c3lc51411k

Li F, Wang X, Liu J, Hu Y, He J (2018) Double-layered microfluidic paper-based device with multiple colorimetric indicators for multiplexed detection of biomolecules. Sens Actuators B Chem 288:266–273. https://doi.org/10.1016/j.snb.2019.02.116

Li Z, Askim JR, Suslick KS (2019a) The optoelectronic nose: colorimetric and fluorometric sensor arrays. Chem. Rev. 119(1):231–292. https://doi.org/10.1021/acs.chemrev.8b00226

Li C, Wang Z, Wang L, Zhang C (2019b) Biosensors for epigenetic biomarkers detection: a review. Biosens Bioelectron 144:111695. https://doi.org/10.1016/j.bios.2019.111695

Li J, Sun Y, Chen C, Sheng T, Liu P, Zhang G (2019c) A smartphone-assisted microfluidic chemistry analyzer using image-based colorimetric assays for multi-index monitoring of diabetes and hyperlipidemia. Anal Chim Acta 1052:105–112. https://doi.org/10.1016/j.aca.2018.11.025

Malon RSP, Chua KY, Wicaksono DHB, Córcoles EP (2014) Cotton fabric-based electrochemical device for lactate measurement in saliva. Analyst 139(12):3009–3016. https://doi.org/10.1039/c4an00201f

Mao H, Zuo Z, Yang N, Huang JS, Yan Y (2017) A microfluidic colorimetric biosensor for chlorpyrifos determination based on peroxidase-like CuFe2O4/GQDs magnetic nanoparticles. J Residuals Sci Technol 14(1):255–269. https://doi.org/10.12783/issn.1544-8053/14/1/30

Matějovský L, Pitschmann V (2018) New carrier made from glass nanofibres for the colorimetric biosensor of cholinesterase inhibitors. Biosensors 8(2). https://doi.org/10.3390/bios8020051

Murdock RC, Shen L, Griffin DK, Kelley-Loughnane N, Papautsky I, Hagen JA (2013) Optimization of a paper-based ELISA for a human performance biomarker. Anal Chem 85(23):11634–11642. https://doi.org/10.1021/ac403040a

Nilghaz A, Lu X (2019) Detection of antibiotic residues in pork using paper-based microfluidic device coupled with filtration and concentration. Anal Chim Acta 1046:163–169. https://doi.org/10.1016/j.aca.2018.09.041

Nilghaz A, Wicaksono DHB, Abdul Majid FA (2011) Batik-inspired wax patterning for cloth-based microfluidic device. In: Proceedings of 2011 2nd international conference on instrumentation control and automation, ICA 2011, pp 82–86. https://doi.org/10.1109/ICA.2011.6130134

Nilghaz A, Bagherbaigi S, Lam CL, Mousavi SM, Córcoles EP, Wicaksono DHB (2015) Multiple semi-quantitative colorimetric assays in compact embeddable microfluidic cloth-based analytical device (μCAD) for effective point-of-care diagnostic. Microfluid. Nanofluidics 19(2):317–333. https://doi.org/10.1007/s10404-015-1545-9

Oncescu V, O'Dell D, Erickson D (2013) Smartphone based health accessory for colorimetric detection of biomarkers in sweat and saliva. Lab Chip 13(16):3232–3238. https://doi.org/10.1039/c3lc50431j

Parikesit GOF et al (2012) Textile-based microfluidics: modulated wetting, mixing, sorting, and energy harvesting. J Text Inst 103(10):1077–1087. https://doi.org/10.1080/00405000.2012.660756

Park BH et al (2017) An integrated rotary microfluidic system with DNA extraction, loop-mediated isothermal amplification, and lateral flow strip based detection for point-of-care pathogen diagnostics. Biosens Bioelectron 91:334–340, 2017. https://doi.org/10.1016/j.bios.2016.11.063

Park CH, Kang YK, Im SS (2004) Biodegradability of cellulose fabrics. J Appl Polym Sci 94(1):248–253. https://doi.org/10.1002/app.20879

Pitschmann V, Matějovský L, Lobotka M, Dědič J, Urban M, Dymák M (2018) Modified biosensor for cholinesterase inhibitors with guinea green b as the color indicator. Biosensors 8(3):81. https://doi.org/10.3390/bios8030081

Reches M, Mirica KA, Dasgupta R, Dickey MD, Butte MJ, Whitesides GM (2010) Thread as a matrix for biomedical assays. ACS Appl Mater Interf 2(6):1722–1728. https://doi.org/10.1021/am1002266

Ricci RW, Ditzler MA, Nestor LP (1994) Discovering the Beer-Lambert law. J Chem Educ 71(11):983–985. https://doi.org/10.1021/ed071p983

Sanjay ST, Dou M, Sun J, Li X (2016) A paper/polymer hybrid microfluidic microplate for rapid quantitative detection of multiple disease biomarkers. Sci Rep 6:1–10. https://doi.org/10.1038/srep30474

Sapsford KE, Francis J, Sun S, Kostov Y, Rasooly A (2009) Miniaturized 96-well ELISA chips for staphylococcal enterotoxin B detection using portable colorimetric detector. Anal Bioanal Chem 394(2):499–505. https://doi.org/10.1007/s00216-009-2730-z

Sayad A, Ibrahim F, Mukim Uddin S, Cho J, Madou M, Thong KL (2017) A microdevice for rapid, monoplex and colorimetric detection of foodborne pathogens using a centrifugal microfluidic platform. Biosens Bioelectron 100:96–104. https://doi.org/10.1016/j.bios.2017.08.060

Sun S, Yang M, Kostov Y, Rasooly A (2010) ELISA-LOC: Lab-on-a-chip for enzyme-linked immunodetection. Lab Chip 10(16):2093–2100. https://doi.org/10.1039/c003994b

Teengam P, Siangproh W, Tuantranont A, Vilaivan T, Chailapakul O, Henry CS (2017) Multiplex paper-based colorimetric DNA sensor using pyrrolidinyl peptide nucleic acid-induced AgNPs aggregation for detecting MERS-CoV, MTB, and HPV Oligonucleotides. Anal Chem 89(10):5428–5435. https://doi.org/10.1021/acs.analchem.7b00255

Wang Y et al (2016) Smartphone spectrometer for colorimetric biosensing. Analyst 141(11):3233–3238. https://doi.org/10.1039/c5an02508g

Wilson H (2013) An approach to chemical analysis: its development and practice

Zhang C, Bailey DP, Suslick KS (2006) Colorimetric sensor arrays for the analysis of beers: a feasibility study. https://doi.org/10.1021/jf060110a

Zheng L, Cai G, Wang S, Liao M, Li Y, Lin J (2019) A microfluidic colorimetric biosensor for rapid detection of Escherichia coli O157:H7 using gold nanoparticle aggregation and smart phone imaging. Biosens Bioelectron 124–125:143–149. https://doi.org/10.1016/j.bios.2018.10.006

Chapter 3
Bio-microelectromechanical Systems (BioMEMS) in Bio-sensing Applications-Fluorescence Detection Strategies

Luis Acosta-Soto and Samira Hosseini

3.1 Introduction

The phenomenon of light emission upon a molecule absorbing electromagnetic energy is called fluorescence (Li et al. 2019). Fluorescence sensing is based on the theory of the target analytes mediated fluorescence quenching ("turn-off") or fluorescence enhancement ("turn-on") (Zhang 2019). Fluorescence biosensors operate based on the measurement of the fluorescence intensity. BioMEMS biosensors are recognized for their portability, excellent sensitivity, rapid analysis, and outstanding selectivity and have been extensively applied for the detection of different biomarkers (Li et al. 2019; Zhang 2019). Fluorescence detection method after colorimetric is the most commonly used spectroscopic method compared to other optical sensing strategies. For data analysis, fluorescence measurements relies on a variety of parameters, including fluorescence intensity, anisotropy, lifetime, emission and excitation spectra, fluorescence decay, and quantum yield (Li et al. 2019).

Fluorescence biosensors can help researchers study and analyze complex chemical processes within cells by incorporation of specific substances into the host cells (Ali et al. 2017). In comparison to the colorimetric detection strategy, fluorescence detection offers higher capacity of anti-inference, and higher sensitivity (Zhang 2019). Fluorescence detection, however, suffers from certain shortcomings including bleaching, light sensitivity, and background noise in signal readout.

L. Acosta-Soto · S. Hosseini (✉)
School of Engineering and Sciences, Tecnologico de Monterrey, Monterrey, Mexico
e-mail: samira.hosseini@tec.mx

S. Hosseini et al., *BioMEMS*, Lecture Notes in Bioengineering,
https://doi.org/10.1007/978-981-15-6382-9_3

3.2 Fluorescence Detection Strategy

Fluorescence is the process of the emission of light by a molecule or material called fluorophore after initial electron excitation in a light-absorption process. After excitation, the fluorophore temporarily retains its activity; this period is called fluorescence lifetime. The fluorophore returns to its original state of energy from that of excited state and the fluorescence emission can be observed with lower energy than the excitation. The fluorescence lifetime depends on the fluorophore and its interactions with the environment (Lakowicz and Lakowicz 1999; Baldini 2009).

Fluorescence detection is a suitable strategy for the development of biosensors since several parameters of fluorescence emission can be measured and recorded. These include fluorescence intensity, fluorescence emission spectrum, fluorescence excitation spectrum, emission anisotropy, and fluorescence lifetime (Stenken 2009). These parameters can be determined as a function of excitation and emission wavelengths. They can be used to detect the presence of different biomarkers and therefore there is an increasing demand for fluorescent sensors with fast response, high sensitivity, high selectivity, portability, and ability to perform real-time analysis. The latest examples of fluorescent BioMEMS biosensors are discussed in Table 3.1.

3.3 Recent Advances of Fluorescence Detection in Paper-Based BioMEMS

Biosensors based on fluorescence detection are designed and fabricated for a versatile class of platforms (Tiwari et al. 2017, 181–190). Paper-based biosensors offer certain challenges alongside and benefits when implementing fluorescence detection. Paper-based devices commonly rely on capillary forces to drive sample and reactants along a path traced for desired interactions. Moreover, these devices have the characteristic of being distinctly simple to operate, analyze, and dispose.

PADs have shown to have been extensively used as detection tools or as parts of more complex devices that may carry out multi-step processing pertaining the sample volume to the micro scale (Table 3.1) (Rosa et al. 2014; Zhang et al. 2015; Sonobe 2019)). As it was mentioned, μPAD operate based upon the capillary forces, offering cost-effective replacements for plastic and glass materials that are typically used for microfabrication. Through capillary forces, the fluids can be directed towards a specific direction by modifying the paper with hydrophobic barriers and patterns that can prevent or allow flow of substrate in highly specific directions. Capillary forces have the potential to drive fluid in a targeted direction without the need for pumps, syringes, or even rotating platforms. Paper are also benefited from the geometry of interwoven fiber network that provide high surface area for enhanced affinity towards analytes, thus facilitating detection. The higher surface area of paper materials allows stronger detection signal and lower limit of detection (LOD). Some of the latest

Table 3.1 Recent BioMEMS platforms for fluorescent detection: Type of the platform, main components, fabrication strategy, mechanism of operation

BioMEMS platform	Main components	Fabrication strategy	Mechanisms of operation	Detected analyte	Specifics	Ref
μPAD	• ZnO-NRs • WFP	The paper device was fabricated by treating WFP with a standard hydrothermal process to grow ZnO-NRs. Using salinization chemistry, capture antibodies were bound to the surface of the treated paper	Functionalized paper was exposed to sample flow to promote analyte-surface interaction followed by exposure to HRP conjugated secondary antibodies for the final readout	Cardiac Myoglobin	The μPAD can be used both as a passive pre-concentrator and an analytical device itself	Tiwari et al. (2017)
	• WFP • CBM-ZZ • IgG-FITC	Micro-channels are wax printed onto the surface of the WFP and then melted. WFP was exposed to CBM-ZZ. IgG-FITC was deposited on to the paper for subsequent detection	The analyte sample was directed in the paper by capillary forces through different analysis stages for biorecognition	Biotinylated-DNA	The final test performed showed little to no detection of analyte despite its known presence, leaving room for improvement of the device	Rosa et al. (2014)

(continued)

Table 3.1 (continued)

BioMEMS platform	Main components	Fabrication strategy	Mechanisms of operation	Detected analyte	Specifics	Ref
	• WFP • $TbCl_3$	The channels and corresponding grading scale were printed onto paper. Later, $TbCl_3$ was deposited on top of the paper to be absorbed	The tear samples were deposited onto the device, and after pre-defined lapse of time the device was tested for fluorescent signals for quantification of lactoferrin concentration	Lactoferrin	An antibody-free method for detection of lactoferrin in tear fluid was experimentally validated	Sonobe (2019)
	• WFP • Geometric radial pattern	The pattern was designed via graphic-design software, and later was wax printed onto WFP using a heating/melting mechanism	Sample was deposited into center port and traveled across the platform by capillary forces towards the analytical zones	• Ag • Hg • Aminoglycans of antibiotics	Though intended to interact with only one antibiotic, the similarity between different antibiotics from this family allows detection of multiple analytes of interest	Zhang et al. (2015)

(continued)

Table 3.1 (continued)

BioMEMS platform	Main components	Fabrication strategy	Mechanisms of operation	Detected analyte	Specifics	Ref
OLED hybrid integrated polymer microfluidic biosensor	• OLED • PDMS microfluidic chip • Inlets and outlets • Rinsing ports	Pre-PDMS was prepared by curing agent prior to casting over metal master-mold, which was later cured in an oven at 90 °C, and then bound to a second PDMS sheet. Moreover, ITO was lithographically patterned and treated by oxygen plasma before being subjected to a vacuum chamber for organic material evaporation	In sterilized PDMS chip, fluids were guided by suction towards mixing chamber. General sandwich immunoassay was carried out to trap the analyte and tag it. To obtain a fluorescent response, the OLED was activated with a 9 V variable DC voltage source, and response was measured through a spectrometer	Donkey anti-sheep IgG	A minimum emission response was detected at a concentration 5 times lower than that of original sample (2 μg/mL) If properly sterilized with water and iso-propyl alcohol (IPA), the device can be used up to 3 times	Acharya et al. (2015)

(continued)

Table 3.1 (continued)

BioMEMS platform	Main components	Fabrication strategy	Mechanisms of operation	Detected analyte	Specifics	Ref
3D microfluidic device with embedded micro-ball lenses array	• 3D multilayer PDMS • Solid micro-ball lenses	A photolithography process was used to fabricate a SU-8 post array on a silicon wafer, which was used to cast PDMS microwells following a soft lithography protocol. Glass microspheres were spread on the surface of the well array and swept into them	A high-power laser diode provided fluorescence excitation on cells through a micro-ball lens array. Multicolor fluorescence emission from cells were collected by the same micro-lens array and imaged by a high-speed CMOS camera through telescope optics	Live Mammalian cells	The fluid flow should be centered in order to achieve optimal uniformity in signal detection	Fan et al. (2013)

(continued)

Table 3.1 (continued)

BioMEMS platform	Main components	Fabrication strategy	Mechanisms of operation	Detected analyte	Specifics	Ref
Lab-on-chip	• Monolith column • LIF system • Inlets	Hot embossing was used to transfer channel features into the propylene microdevices with the help of CNC aluminum masters, while the holes were drilled for inlets and outlets. The monoliths were formed by a single-step method by filling a device with polymerization mixture and photo initiator and exposing to UV light	A lysate sample was directed through the column monolith. The porous allowed selectivity to the target DNA. After a rinse step, captured DNA was released by heating the monolith above the melting point of the DNA	dsDNA from the carbapenemase gene	The lowest detected concentration of KPC amplicons was 100 pM using this LIF system. The capture efficiency of the target was 86%, while its labelling was done with 97% accuracy	Knob et al. (2018)

(continued)

Table 3.1 (continued)

BioMEMS platform	Main components	Fabrication strategy	Mechanisms of operation	Detected analyte	Specifics	Ref
	• CELL chips Oval-shaped chambers • CGG • Inlets • MIX chip	The chips were fabricated by soft-lithography and wet etching. Each resulting PDMS substrate was bonded to a glass surface by exposing the samples to air plasma	HeLa cells were cultured in 5 different CELL chips. The CELL chips are disconnected from the CGG chip to be placed within incubator. The MIX chip was connected to the CELL chip to let the probe solution interact with the cells	PS beads as model analyte	Proposed platform reduces the time necessary for the assay by ~40%	Montón et al. (2017)

(continued)

Table 3.1 (continued)

BioMEMS platform	Main components	Fabrication strategy	Mechanisms of operation	Detected analyte	Specifics	Ref
	• PDMS • Micro-spheres	The PDMS chip was fabricated through soft lithography bound to glass slides using a plasma cleaner. The polystyrene micro-spheres were exposed to streptavidin, and the resulting conjugates were bound to a primary antibody for initial entrapment of insulin or insulin analog	A sample of T1D patients' blood was taken and pumped through the inlet of the device. As the sample flows through the channel, it is intersected with functionalized micro-spheres and subsequently conjugated with secondary antibody fluorescently tagged. The mixture travelled through outlet channel to be imaged for the final readout	Insulin analogs	The device accurately detected the insulin levels within 30s of injecting sample into the chip	Cohen (2017)

(continued)

Table 3.1 (continued)

BioMEMS platform	Main components	Fabrication strategy	Mechanisms of operation	Detected analyte	Specifics	Ref
	• PDMS chip • Bacterial trap Poly-Si-photogate	Through soft lithography, the PDMS chip was made from SU-8 mold and subjected to oxygen plasma to bind to lid. Channel and chamber geometry were defined accordingly to accommodate the micro-beads. Si-photo sensor was placed directly below the chip	Bacteria solution was pumped into the chip along with microbeads that helped trapping them within the setup fluorescence detection. The sample was excited through a UV light source, which was turned off to allow the Si-photogate to detect the fluorescent signals	L. pneumophila cells	This method offered a highly sensitive and timely option to identify bacterial presence without the need for optical sensors of more complex equipment for detection	Onishi (2017)
Electro-mechanical sensor	• 3D printed case • Microcontroller • Circuitry • Sample loading port	A ready-to-use biosensor that integrates a microcontroller, and corresponding circuitry to control a LED for excitation	Prior to loading the analyte solution, the device was calibrated with known concentrations for efficiently estimating the biomass composition of unknown sample	• Chlorophyll A and B • Phycocyanin	The error rate of the device was rather high (2–16%)	Shin et al. (2018)

(continued)

Table 3.1 (continued)

BioMEMS platform	Main components	Fabrication strategy	Mechanisms of operation	Detected analyte	Specifics	Ref
Alternative BioMEMS	• Micro-bead • RhB • FITC	Micro-beads were first incubated in alcohol solution to cause swelling. RhB penetrated the network and stained the beads. The beads were washed in deionized water to close the pores. Finally, it was submerged in a saturated solution containing FITC for staining and imaging	Upon excitation, the two fluorophores emitted different responses, one dependent solely on temperature (rhB), and another both on temperature and pH (FITC). This difference was used to identify them	Amino-PS beads based on temperature and pH	Sensor offered a great accuracy at 0.1 °C and 0.2 pH	Liu et al. (2014)

(continued)

Table 3.1 (continued)

BioMEMS platform	Main components	Fabrication strategy	Mechanisms of operation	Detected analyte	Specifics	Ref
	• Pt imprinted micro-motor • PC film • Pt • Ni	A PC film were Pt-deposited which was followed by coating with Ni, and a final layer of Pt. The surface of the cone-like structures which were integrated within the PC film was magnetically imprinted with phycocyanin molecules to leave exposed sites for binding	The device was placed in sea water. A combustible for propulsion was administered to the medium to initiate the motor's motion. The device can was guided with magnets in desired directions and was intentionally left in the sample prior to readout	Phycocyanin	This study offers a novel method for biorecognition that relies on magnetic force for locating the device	Zhang et al. (2013)

(continued)

Table 3.1 (continued)

BioMEMS platform	Main components	Fabrication strategy	Mechanisms of operation	Detected analyte	Specifics	Ref
	• Functionalized QDs • Organic quencher (BHQ2)	The QDs were conjugated to DNA strands designed to bind to analyte of interest	The sample was prepared by exposure to the analyte marker. In the presence of the analyte is present, will interact with the QD and quench the fluorescent signal. The mechanism of operation was based on Förster resonance energy transfer (FRET)	• DNA • MicroRNA	The DNA-QDs is benefited with excellent optical properties of QDs while using tailored properties of DNA for detection hence act as nano-sensors for nucleic acid detection	Su et al.(2014)
	• AuNC • FITC • PBS	AuNC was exposed to PBS to form PBA-AuNC. FITC was conjugated to this coupling complex and FA was subsequently incorporated to improve cellular membrane permeability	The complex of FA-FITC-BSA-AuNC demonstrates a high affinity towards the folate acceptor proteins on the cytoplasmic membranes of Hela cells. Once it has crossed the membrane, it can be used to monitor changes in pH as a function of the ratio of the stable AuNC emission in contract to that of FITC	Intracellular pH	The AuNC is highly permeable to the cytoplasmic membrane of cancerous cells	Ding and Tian (2014)

(continued)

Table 3.1 (continued)

BioMEMS platform	Main components	Fabrication strategy	Mechanisms of operation	Detected analyte	Specifics	Ref
	• Perceval sensor • PCR	The strategy for development of PercevalHR was based upon mutating the original Perceval sensor around the GlnK nucleotide-binding site	Relative concentrations of ATP to ADP within cells was used to monitor the metabolic activity of cells. Once the molecule has permeated the cytoplasmic membrane, fluorometric response to different excitation levels were monitored to determine the ATP:ADP ratio	Metabolic activity of cells	Although it brings a significant improvement in detection range over the first iteration, it still falls short of the total range that can be found within cells	Tantama et al. (2013)

The analyte of interest, and the advantages and disadvantages of each platform are presented in this table
Annexin V (Ann-V); Camptothecin (CAMPT); Computer numerical control (CNC); Direct current (DC); Double-strand DNA (dsDNA); Concentration gradient generating chip (CGG); Fluorescein isothiocyanate (FITC); Idium tin oxide (ITO); Immuno-globulin G (IgG); Iso-propyl Alcohol (IPA); Lab on chip (LOC); Laser-Induced Fluorescence (LIF); Microfluidic paper based analytical device (μPAD); Organic electroluminescent diode (OLED); Phosphatidyl-serine (Ps); Polydimethylsiloxane (PDMS); Polycarbonate (PC); Polymerase chain reaction (PCR); Quantum dots (QD); Terbium (Tb); Type 1 diabetes (T1D); Whatman filter paper (WFP); Zinc oxide nano-rods (Zn-NRs)

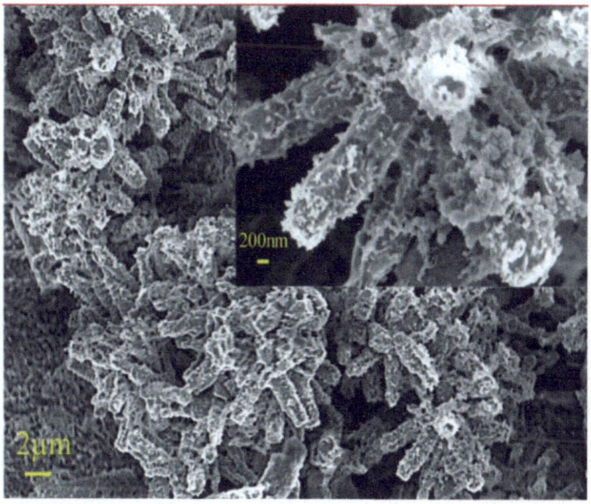

Fig. 3.1 WFP after functionalization; inset shows close-up of nano-rods after antibody immobilization (Tiwari et al. 2017)

examples of the paper-based BioMEMS used for fluorescence detection are provided in this chapter.

Tiwari et al. (2017) investigated the performance of Zinc oxide nanorods functionalized paper (ZnO-NRs-WFP) compared to regular Whatman filter paper (WFP) in the capture, detection, and release capabilities of the cardiac myoglobin. WFP was functionalized with nanorods (Fig. 3.1) that increased the surface area available to bind to the surface primary antibodies. With these paper alterations, the device showed a threefold improvement when compared to the control (non-functionalized) WFP. Although enzyme-linked immunosorbent assay (ELISA) protocol was used to demonstrate sensitivity and performance of the device through fluorescent detection of analyte posterior to exposure to secondary antibody, the results adduced the capability of molecular trapping, justifying future use as a preconcentration stage integrated into a multi-stage biosensor.

One of the main challenges of the paper materials for their use in biosensing is the fact that they commonly require functionalization to better interact with biomolecules. μPADs can be exposed to the biomolecules without functionalization as well (physical attachment). However, through many washing steps involved in bio-assays, these molecules may leach out of the fibrous structure of paper which expectedly results in undesirable detection outcomes. To avoid this, Rosa et al. (2014) investigated the use of the third family of carbohydrate binding module (CBM) of the *Clostridium Thermocellum*. This CBM has a high specificity towards cellulose and is therefore used as a part of a larger conjugate in order to bind the ZZ-domain of the staphylococcal protein A to the paper fibers. Once this interaction took place, the ZZ-domain from its free end can be used to bind to an IgG antibody. The IgG, in turn, can bind to and detect the presence of biotin. For that reason, if a DNA strand is biotinylated and fluorescently tagged, the presence of a fluorescent response in the paper would attest to the device's effectiveness in detection. The study demonstrated that its mechanism of operation provides a reliable means for detection of analyte even though the signal was not comparable to that of EFP. This shortcoming was due to the lack of control over the orientation of fibers within the paper structure, resulting in unfavorable bonds with orientations that did not lend themselves to predetermined bindings hence dramatically decreasing the sensitivity of the device.

Dry eye disease (DED) is a condition that affects patients by causing complications in the production of lactoferrin, a chemical secreted in tears that protects the eye from an array of threats. This pathology is often diagnosed after a number of tests including ELISA. The μPAD proposed by Sonobe et al. (2019) operates on the basis of detection of lactoferrin in tear fluid. For fabrication of this device, WFP was embedded with terbium (Tb). Tb^{3+} reacts to lactoferrin and its conformation provides a fluorescent response to an excitation source. This response can be measured in terms of relative intensity, and, along with the linear design of the device, the height to which the column shows a fluorescent stain prior to excitation is indicative of the concentration of lactoferrin within the sample. Thus, an antibody-free method for detection of lactoferrin in tear fluid was proposed and experimentally validated (Sonobe 2019).

One of the main objectives of BioMEMS biosensors is to optimize the assay time. One of the methods for achieving a shorter analysis time is simultaneous screening of different analytes from a single sample. In the device proposed by Zhang et al. (2015) the simultaneous detection of three different contaminants in food was carried out. The geometry of the device allowed a central sample deposition from which the sample was guided towards several channels with already functionalized biomolecules targeted at different analytes. Figure 3.2 shows the scheme of the device. The surface of WFP was treated with graphene oxide, and the designed

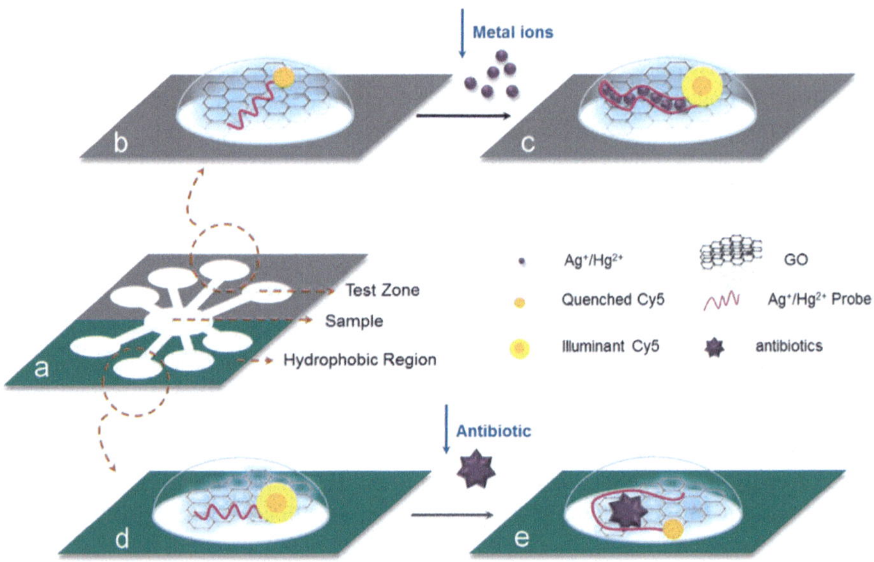

Fig. 3.2 Schematic of the final μPAD (Zhang et al. 2015)

geometry was created on the paper by wax printing. A half of the device was dedicated to the detection of metallic ions including Hg^{2+} and Ag^{+} via functionalization with ss-DNA strand that quenches the fluorescence of deposited Cy5 marker. Upon contact with the metallic ions, Hg^{2+} and Ag^{+} reacted with the thymine and cytosine bases respectively, which freed the Cy5 marker and allowed a fluorescent response to be read out. The second half of the device was devoted to the sensing of antibiotic residues. The method of detection in this other half was almost opposite as shown in Fig. 3.2. The interaction between the antibiotic residue and the graphene oxide surface, with the fluorescent probe, resulted in fluorescent quenching. Thus, the detection of this third analyte was measured by a decrease in the fluorescence signal. The device has shown a high level of specificity for the detection of the metal ions. However, it showed cross-compatibility between aminoglycoside antibiotics due to the similarly arranged amino groups that could react with the epoxy groups of the graphene oxide surface as well hence resulting in less reliable outcomes.

3.4 Microfluidic BioMEMS Recent Advances of Fluorescence Detection in Microfluidic BioMEMS

3.4.1 Recent Advances of Fluorescence Detection in Lab-On-Chip (LOC) Devices

Through the combination of biological assays such as ELISA with microfluidic technologies, it is possible to design biosensors that operate on minimal sample volumes and offer accurate detection outcomes, while being portable, and often reusable (Acharya et al. 2015; Fan et al. 2013; Knob et al. 2018; Montón et al. 2017; Onishi 2017). In this chapter some of the latest examples of the microfluidic BioMEMS for fluorescent detection are as provided.

The organic electroluminescent diode (OLED) has been studied as an economic and easy to manufacture source of light (energy). It is an attractive option not only for the possibility of reducing production costs, but also for the high degree of specificity it offers with respect to the light wavelength it emits. A device that integrates OLED in a microfluidic device was reported by Acharya et al. (2015). The proposed LOC (Fig. 3.3) used an AlQ3 OLED to activate fluorescent dye *Alexafluor 488* as part of a fluorescent assay for bio-optical detection of the antigens. The target antigen was anti-sheep IgGs, which were fluorescently tagged and were detected via a sandwich immunoassay after conjugating with monoclonal antibodies (Fig. 3.4). A common issue when designing such devices is the interference of the light source with the fluorescence as it may result in bleaching. To avoid such problem, a filter can be used to allow only certain wavelength of the emitted light to reach the sample and sensor. The excitation peak of the fluorophore should then be lower than the emission peak, thus allowing the tag's fluorescence to accurately reflect the presence of the antigen. Another concern regarding the OLED source is the inevitable heating of the

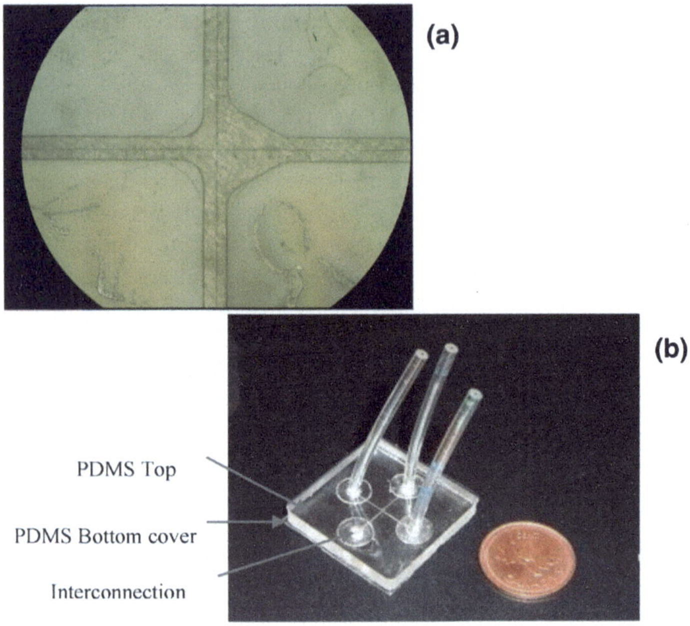

Fig. 3.3 Close-up of the device's chamber is shown on top (**a**); Below, the LOC from afar with the inlet an outlet channels labeled (**b**) (Acharya et al. 2015)

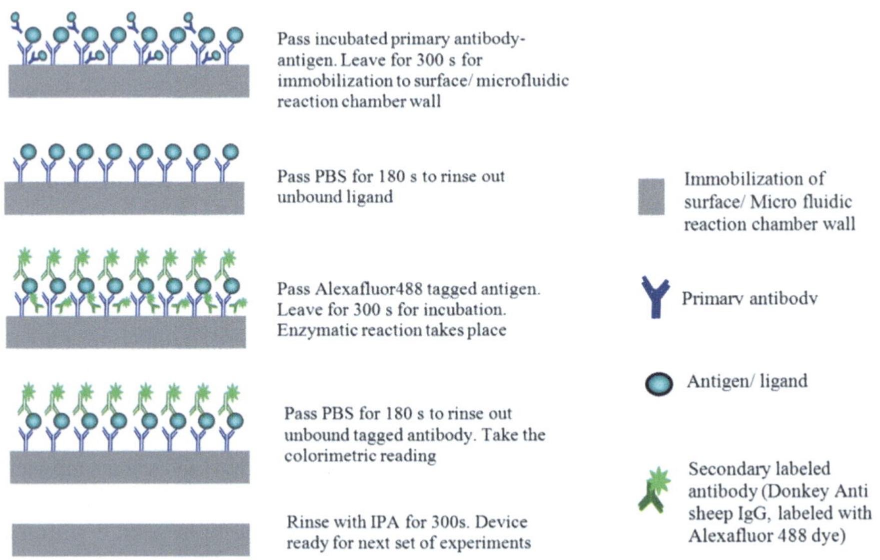

Fig. 3.4 An assay carried out for detection of anti-sheep IgG on AlQ3-OLED-LoC (Acharya et al. 2015)

components. To keep the effect of this increased temperature at negligible levels, the OLED was fed 9 V through a variable DC voltage source. Lower levels of LOD was achieved when the antibody was diluted 5 times the original concentration (2 μg/mL), that corresponded to a 1:1 ratio between the antigen and the secondary antibody. No relevant fluorescence was detected by the spectrometer after this point. The saturated sample with high concentration of biomolecules does not promote efficient binding between antigens and antibodies due to the possibility of steric repulsion. Despite the advantages of fluorescence-based biosensors utilizing an OLED, certain applications have design requirements that may place this as a suboptimal option.

In the case of multiple single sensing, various LOC configurations offered excellent platforms (Table 3.1). An example is a device that can screen up to 188,800 cells per second by profiling cells into 32 different channels (Fig. 3.5) that are simultaneously imaged (Fan et al. 2013). To achieve such high rate of imaging without the risk of false positive signal, the flow through the sample channels must be carefully focused. It was observed that the use of embedded micro-ball lenses (Fig. 3.5a) focused the analytes in the *x*–*y* plane at variable heights within the channel geometry. The spherical chromatic aberration of the lenses naturally magnified the fluorescent signals emitted by the assay, so long as the signal was aligned directly above the lens, which gave the LOC a high sensitivity. However, since the light entered the micro-ball at once, if the procedure, for instance, demanded a cell to be tagged with two fluorophores, the colors would appear blended into a third tone on the final image. To separate the light into its original components, before reaching the camera lens, the light has traveled through a prism that refracted the emissions (Fig. 3.5b). By using

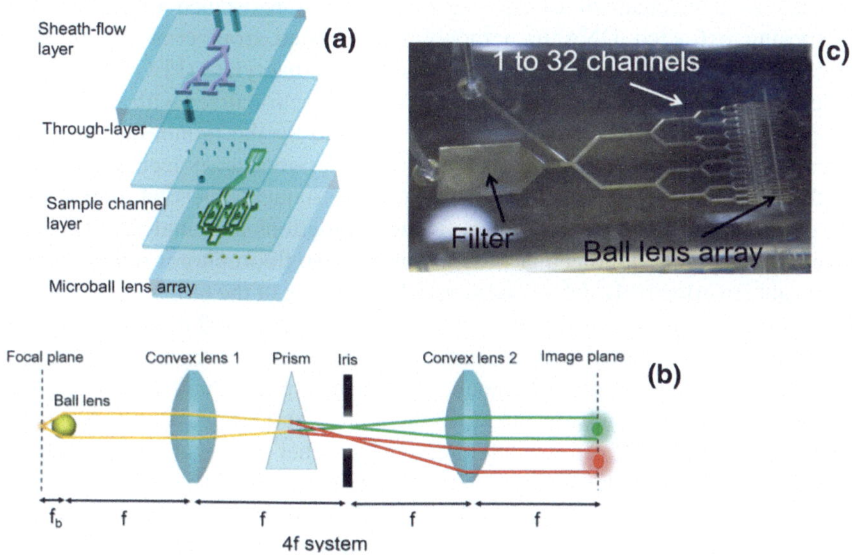

Fig. 3.5 **a** Shows the structure and build of the PDMS device **b** Portrays the fluorescence signal's trajectory **c** Shows final device (Fan et al. 2013)

this strategy, the device was tested with fluorescent dye *Lucifer yellow* in presence of the emission filter *Chroma 59,004* with transmission windows at both 500–535 nm and 570–620 nm resulting in satisfactory imaging and detection of the two windows. The smaller range corresponds to the green emission and the larger one to the red emission. An additional reason that this setup is a viable option for large-scale cells flow cytometry is the optimization of signal to the background and the source noise. The utilization of micro-ball lenses has improved this ratio 18 times against detection the systems without micro-ball lenses.

The objective of flow cytometry studies is usually to study live cells, rather than dyes. Fan et al. (2013) achieved detection of two cells (HeLa and Ramos cells) with different fluorescent tags along with accurate readout of respective proportions in the sample supplied to the device. Multiple and simultaneous imaging, however, is a demanding process, particularly when large volumes and fast results are needed. In any case, the sample must be initially tagged, which is not a simple task when the analyte is a strand of a genetic material.

One of the applications of LOC devices is the detection of antibiotic resistance in sepsis-related scenarios (Knob et al. 2018). In this case, specific DNA strands must be located and tagged for further analysis of a sample. This preparation requires the lysis of the sample bacteria cells, obtained through prior separation from the patient's cells. Once the genetic material is available, it is possible to employ various strategies to bind molecular markers to a given strand. Recently, a monolith within a polypropylene (PP) was used in a microfluidic device with specific functionalization to trap DNA sequences. In their study, Knob et al. (2018) immobilized genes related to Klebsiella pneumoniae carbapenemase (KPC). This was used as a mechanism to analyze the bacterial resistance to carbapenems by exposing extracted DNA to a monolith fabricated from crosslinkers exclusively. These crosslinkers were polyethylene-glycol diacrylate (PEGDA) and 3,4-Ethylenedioxy-N-methylamphetamine (MDMHA) functionalized with a tailored 90-mer specific to the DNA of interest. Once captured by the monolith, the genetic material was then exposed to hybridization probes designed as molecular beacons to bind the fluorophore to the target. For this step, the sequence needed to bind to the KPC genes was determined through software simulations and analysis. The study showed that a modified molecular beacon that has two fluorophores rather than just one was able to effectively detect and signal the presence of the target DNA. This has validated the use and functionality of the single-step fabricated monolith proposed for capture, labeling, and eluting of analyte.

Our understanding of the efficiency of molecular probes and dyes as fluorescent emitters has led the advancements in the field to another class of fluorophores, quantum dots (QD). QDs offer an alternative with higher photostability, availability, and adaptability for specific applications. QDs were applied in the detection of carcinoma cell apoptosis (Montón et al. 2017), as described in Table 3.1. When cells enter the state of apoptosis, it is common to observe the translocation of phosphatidylserine (Ps) from the inner layer of the cellular membrane to the exterior one. This presents a unique possibility to identify this process of cellular death by tracking Ps translocation through QD tagging. By binding a QD to the apoptotic cell, annexin V

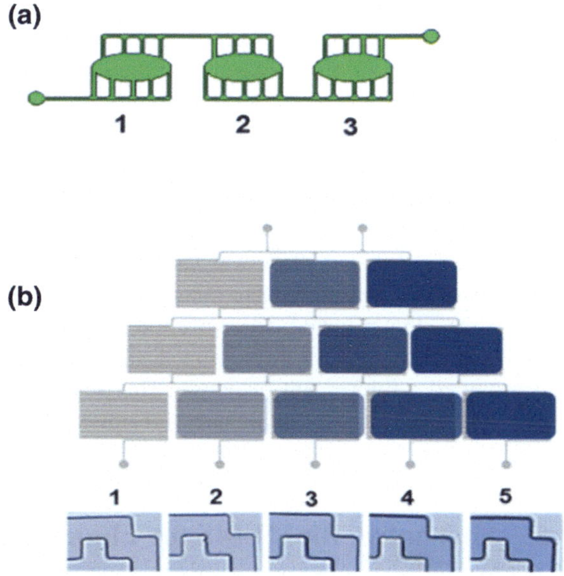

Fig. 3.6 **a** CELL chip **b** Concentration gradient generation chip exemplifying the 5 concentrations of CAMPT used by using a dye (Montón et al. 2017)

(Ann-V) can be conjugated to the QD. Ann-V has a high affinity towards Ps, acting as intermediary link between QDs and the apoptotic cells.

In the microfluidic device investigated by Montón et. al. (2017), two stages are proposed, each with their respective LOC, to cultivate, mark, and identify induction of apoptosis in cancer cells. The CELL chip proposed by the authors had different chambers for cell seeding (Fig. 3.6a). To achieve a monolayer on the CELL chip, samples were incubated for 24 h prior to exposure to camptothecin (CAMPT). Subsequently, the previously conjugated Ann-V-QD (in a separate PDMS device) was pumped into the CELL chip. This procedure was repeated for 5 different concentrations of CAMPT. The results showed a correlation between doses of the cancer treatment chemical and the Ann-V-QD labeled cells, establishing this device as a reliable tool for the bioassays. In the second design in order to induce cellular death, CAMPT was used again, however, it was diluted to a concentration gradient allowing the device to expose cancer cells to different CAMPT doses (Fig. 3.6b).

Modularity grants a high degree of adaptability to assays conducted within LOC devices at the cost of increased complexity. An approach to effectively implement multi-step assays inside a device was reported for determining insulin level and insulin-like factors in type 1 diabetes (T1D) patients (Cohen 2017). The proposed LOC relied on fluid mixing through the induction of transverse flow via serpentine geometries. The sample was introduced to the device via the inlet farthermost from the single outlet of the chip. The sample, once inside the device, was exposed to microspheres previously functionalized with antibodies specific to insulin and insulin-like factors. Taking advantage of the high surface area of spheres, exposure to analyte was magnified, reducing the time needed to achieve satisfactory mixing.

A secondary anti-insulin antibody conjugated with a fluorophore was then added to the system that signaled the presence of the analyte. The PDMS chip showed accurate readout of insulin in blood plasma within 30s without the need for washing of the secondary antibody. This allowed a continuous flow and real time detection within the device.

Although molecular tagging is typically used in fluorescence-based biosensors, some cells have a fluorescent response to UV exposure as well that can be used for detection purposes. In a device designed by Onishi et. al. (2017) the fluorescence of bacteria cells *L. pneumophila* was used to identify its presence within a sample. To trap the bacteria cells, they were mixed with micro-beads that helped form a barrier, or "stopper", that allowed concentration of bacteria cells to increase dramatically around the stopper. Once immobilized, the bacteria were exposed to UV radiation that provided a fluorescent response. This sensor enabled the detection of natural fluorescent bacteria without the need for optical or elaborate sensors and cameras.

3.5 Alternative BioMEMS for Fluorescence Detection

The feasibility of micro-beads as means to monitor changes in both pH and temperature was investigated by Liu et al. (2014). It was found that by using polystyrene (PS) micro-beads in combination with two different fluorescent markers (Rhodamine B and FITC) the fluorescent response of both fluorophores to an excitation source could be processed to obtain information about the medium in which they were submerged (Fig. 3.7). The PS micro-spheres were exposed to alcohol, causing them to swell thus via the open pores, Rhodamine B penetrated and stain the beads. To capture the Rhodamine B particles, the micro-beads were then rinsed in deionized water to shrink back the size hence closing the pores. Afterwards, the surface of the beads was stained with FITC. Since Rhodamine B and FITC have different excitation ranges, high wavelengths were used to elicit a fluorescent response. To demonstrate the detection mechanism, the microbeads were analyzed throughout ranges of both

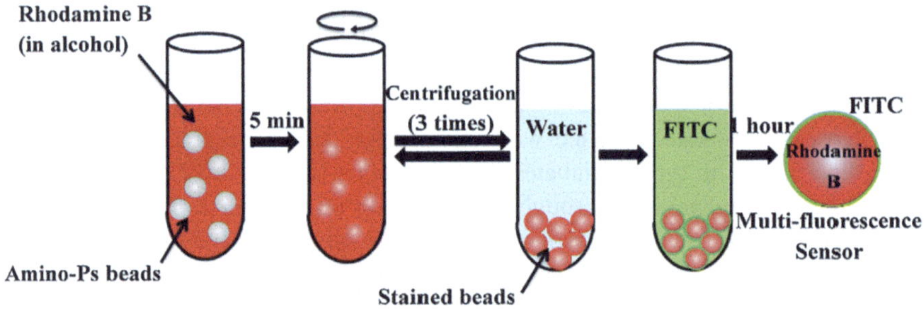

Fig. 3.7 Fabrication process of the micro-beads functionalized with Rhodamine B and FITC (Liu et al. 2014)

temperature (32–38 °C) and pH (5–8). Benefiting from the dependence of Rhodamine B's emission response on temperature, and the independence of it on pH levels, this compound was used for the calibration of the temperature within the medium. This information was then accounted for identifying the temperature dependence of the emission response generated by FITC, in order to accurately to identify the pH and its change in the experiment. The response to cyclic stimulation was observed to be remarkably consistent, as the beads gave the same response for a given configuration of the independent variables several times.

Zhang et al. (2013) reported a biosensor capable of selectively detecting, capturing, and transporting the analyte of interest, phycocyanin, which is closely related to the biomass of certain cyanobacteria (Fig. 3.8). This compound is a natural fluorophore acting as its own marker for detection and is commonly targeted for environmental applications. For fabrication process, a first layer of platinum (Pt) was deposited on a template membrane and then exposed to phycocyanin in an electric field that forced them through the membrane's pores and bound them to the outside of the cone-like structures (Fig. 3.8). Metal ions, Ni^{2+} and Pt^{4+}, were then electrochemically reduced from a solution into previously Pt-backed pores inside a template polycarbonate membrane. Once the air had been removed from the generated structures, the cone-like devices were modified with poly(3,4-ethylenedioxythiopene) (PEDOT)

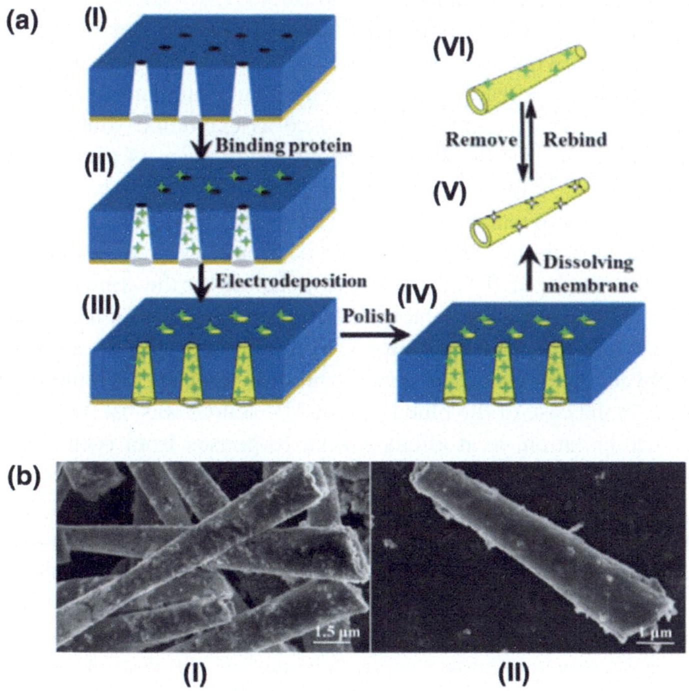

Fig. 3.8 **a** Elaboration of the magnetically imprinted biosensors. **b** Various (I) and single (II) magnetically imprinted cone-shaped sensors (Zhang et al. 2013)

layers that were electrochemically deposited onto the surface of the devices. Subsequently, a single Pt layer was galvanostatically deposited, to enhance the mechanical performance, followed by subsequent depositions of Pt-Ni, solely Ni, and finally Pt, in the same fashion (Fig. 3.8a). The template membrane was washed and sputter coated with gold. As a consequence, this fabrication strategy, the micro-motor was susceptible to magnetic fields, particularly when in accordance to its momentum. It was shown the micro-motor could be manipulated and guided towards specific areas when using sufficiently strong magnets. It is important to note, however, that control over micro-motor's motion depended on the geometry of the cone, as more asymmetrical structures will tend towards erratic spiraling paths given an uneven propulsion. The devices exhibited a targetable movement to target the phycocyanin. The earlier exposures of the cones left available binding sites on the outside surfaces of the cones (imprinted sites) that allowed a relatively quick adsorption of phycocyanin to the surface. Later evaluation in actual seawater showed that the presence of different molecules and compounds did not significantly interfere with the rate of phycocyanin adsorption or the movement and direction control of the micro-motors in the medium.

The detection of cyanobacteria (*Spirulina*) can be done using phycocyanin.

In specific cases, a need arises to simultaneously monitor the presence and ratio of several species within the same space. Such a scenario happens in biofuel production, where a specific ratio of green algae (*Chlorella vulgaris*) to cyanobacteria is desirable. Shin et al. (2018) took advantage of the fact that green algae produce chlorophyll a and chlorophyll b, both of which can produce a fluorescent response to the right stimuli. The proposed device is a ready-to-use biosensor that integrates a microcontroller, corresponding circuitry to control a LED used as excitation sources, along with an amplification circuit for the signal readout. The device is also equipped with a temperature compensating mechanism for the photodetector with an inlet for a vial with the sample to be analyzed and a screen that outputs the data from the test (Fig. 3.9). Unlike most fluorescent devices, this strategy heavily depended on the code that reads the sensors input. The code was responsible for the control and operation of the 3 LEDs that were used to obtain different responses from the three different fluorescent substances within the samples. Through the use of an amber LED, the response of phycocyanin was maximized, while for the two chlorophylls the signal was amplified. In the case of the blue LED, an UV source was used to measure total phytoplankton population, as it elicits similar responses from both *Spirulina* and *Chlorella vulgaris*. The algorithms used to process the information applied concepts of linear algebra analysis to correlate readouts of the three excitation sources and find relative concentrations of the different organisms. The partial least square regression (PLSR) method was used to best estimate the biomass of the different phytoplankton species. The device was calibrated by reading different samples with predetermined concentrations, and manually inserted into the microcontroller, before it can begin to quantitatively estimate unknown values. The device's predictions were within 2–16% of the real values of the respective biomasses. Reducing the background noise can further enhance the accuracy of prediction by this device.

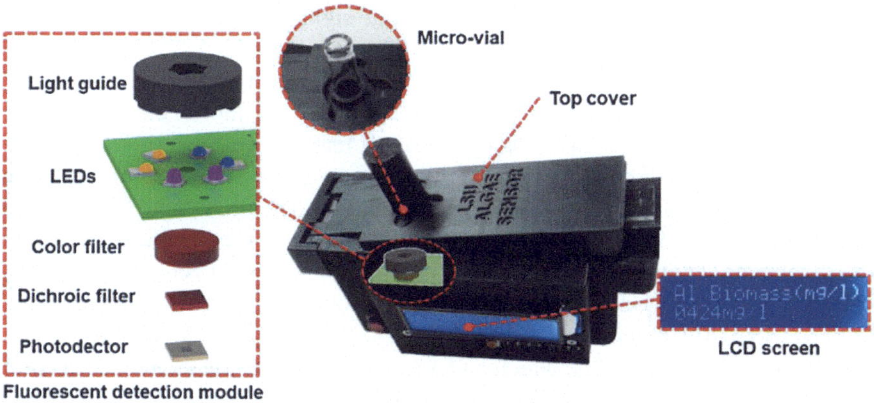

Fig. 3.9 Build of the multiple phytoplankton biosensor (Shin et al. 2018)

The identification of specific strands of genetic material including DNA or micro-RNA (miRNA) is of clinical importance to the detection and diagnostic of several pathologies. A new strategy for detection of such analytes with a high rate of specificity and selectivity was reported by Su et al. (2014). The method involved 3-mercaptopropionic acid (MPA)-capped QDs functionalized with thiolated DNA through ligand-exchange. The two reactants were mixed for several hours at different temperature ranges to achieve complete exchange between the two, resulting in DNA-QD conjugates. Förster resonance energy transfer (FRET) mechanism was used to ensure efficient energy transfer between the QD's and the BHQ_2-DNA (Fig. 3.10).

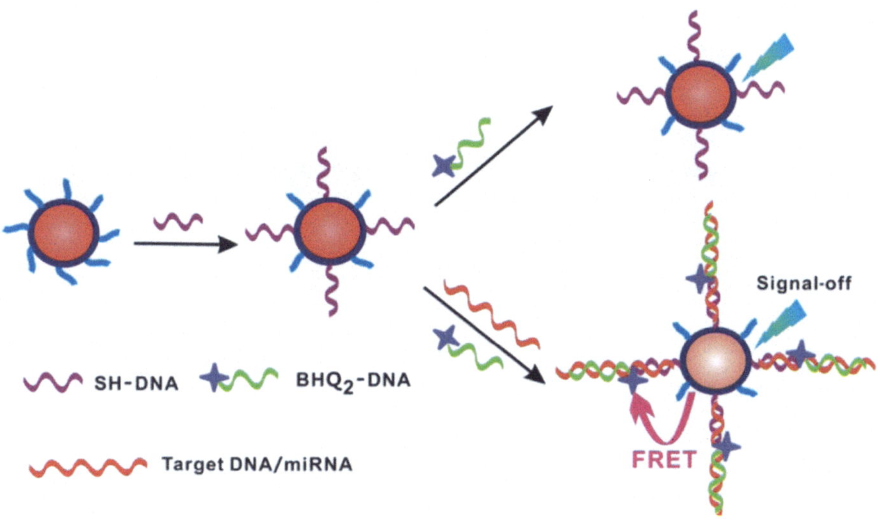

Fig. 3.10 Graphic visualization of FRET mechanism (Su et al. 2014)

The BHQ_2 served as an organic quencher that signals the acquisition of target sequence. The functionalized DNA-QD's exhibited a high degree of specificity in binding to target analyte, while avoiding interactions with other genetic sequences.

The previously alluded capability of FITC to tailor its fluorescent response to excitation depending on the pH of its medium was further investigated by Ding et al. (2014). Gold nano-clusters (AuNC) were encapsulated in bovine serum albumin (BSA) to protect the subjacent AuNC from the medium. The BSA-AuNC was stained FITC. Noteworthy, AuNCs emit a constant fluorescent response specific to its excitation wavelength, giving a static reference point that can be used to compare and calibrate the changes in the FITC response to the surrounding pH levels. Furthermore, the FITC-BSA-AuNC was exposed to folic acid (FA), creating a complex of FA-FITC-BSA-AuNC (Fig. 3.11), which had a high affinity towards the folate acceptor proteins on the cytoplasmic membranes of Hela cells. Due to the stability of the fluorescent response generated by AuNC (F_{AuNC}) at any pH, the ratio F_{FITC}/F_{AuNC} was used to determine the intracellular pH levels. The viability of this new sensor was tested in the presence of various ions typically found in the interior of complex living cells, including human cancer cells. The reported data supported the accuracy of the device, along with a short response time, indicating that treated AuNCs could be used for real time analysis of complex bioassays and biologic interactions.

Relative concentrations of ATP to ADP within cells represent a means to track the metabolic activity cells. Several attempts in fabrication of biosensors capable of crossing the cytoplasmic membrane and monitoring this ratio were made. Tantama et al. (2013) proposed a new version of their previous biosensor Perceval, renamed

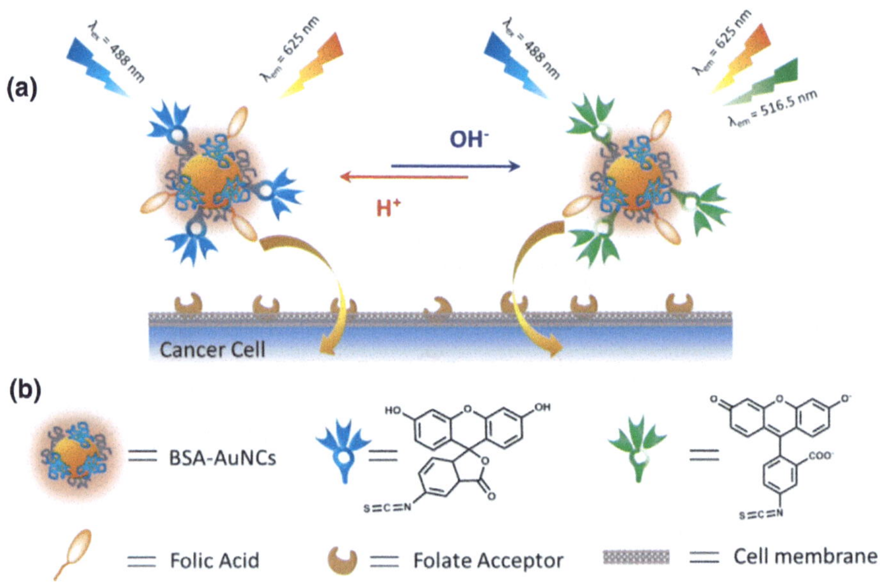

Fig. 3.11 Mechanism of operation of FA-FITC-BSA-AuNCs (Ding and Tian 2014)

as Perceval High Range (PercevalHR). The mechanism of operation in this biosensor relies on its affinity towards both ATP and ADP, while each reacts to different fluorescent wavelengths. ATP produces a fluorescent response for ~500 nm excitation, whereas ADP responds to ~420 nm. Through this difference, a ratio can be established for the relative difference between the responses recorded at one excitation or another, along with a third point (reference) at ~455 nm as an isosbestic fixed point. The biosensor was tested for neurons successfully. While adequate response was recorded to three different stimuli presented to disrupt regular metabolic rate of cells, the measured response was heterogeneous when tested for astrocytes. Such differences were minimized when pH bias was taken into consideration. The authors claimed to have recorded distinct responses for both ATP and ADP including the images recorded by two-photon microscopy. This method, in particular, opens a possibility for simultaneous screening of different metabolic activity within layers of a tissue.

3.6 Summary

Recent advances in biosensors based on fluorescent detection demonstrated that the field of BioMEMS holds great promise for portable and reliable devices capable of performing high throughput bio-assays. Recent studies point towards the versatility of fluorescence detection by employing different fluorophores depending on the materials and reactants involved aimed at optimizing the signal output. The mission of BioMEMS biosensors is to minimize the sample size, while reducing the time and expertise required to carry out these lifesaving assays. To this end, papers, as a class of material that lends an array of properties for integration in various biosensor platforms made one of the greatest candidates for fluorescence detection. Paper can be integrated into different designs of LOCs while it can individually serve as a device itself. Fluorescence detection strategy portrays a series of beneficial features including high sensitivity, and ease of readout. Nonetheless, the fluorescence may also suffer from background signal noise, bleaching, and cross talk effects that impose certain challenges on this widely applied technique (Ramon et al. 2017).

References

Acharya A, Packirisamy M, Izquierdo R (2015) OLED hybrid integrated polymer microfluidic biosensing for point of care testing. Micromachines 6(9):1406–1420. https://doi.org/10.3390/mi6091406

Ali J, Najeeb J, Asim Ali M, Farhan Aslam M, Raza A (2017) Biosensors: their fundamentals, designs, types and most recent impactful applications: a review. J Biosens Bioelectron 08(01). https://doi.org/10.4172/2155-6210.1000235

Baldini F (2009) Alexander P. Demchenko: introduction to fluorescence sensing. Anal Bioanal Chem395(5):1195

Cohen N et al (2017) Microsphere based continuous-flow immunoassay in a microfluidic device for determination of clinically relevant insulin levels. Microchim Acta 184(3):835–841. https://doi.org/10.1007/s00604-017-2072-z

Ding C, Tian Y (2014) Gold nanocluster-based fluorescence biosensor for targeted imaging in cancer cells and ratiometric determination of intracellular pH. Biosens Bioelectron 65:183–190. https://doi.org/10.1016/j.bios.2014.10.034

Fan YJ et al (2013) Three dimensional microfluidics with embedded microball lenses for parallel and high throughput multicolor fluorescence detection. Biomicrofluidics 7(4). https://doi.org/10.1063/1.4818944

Knob R et al (2018) Sequence-specific sepsis-related DNA capture and fluorescent labeling in monoliths prepared by single-step photopolymerization in microfluidic devices. J Chromatogr A 1562:12–18. https://doi.org/10.1016/j.chroma.2018.05.042

Lakowicz JR, Lakowicz JR (1999) "Introduction to Fluorescence", in *Principles of Fluorescence Spectroscopy*. Springer, US, pp 1–23

Liu H, Maruyama H, Masuda T, Honda A, Arai F (2014) Multi-fluorescent micro-sensor for accurate measurement of pH and temperature variations in micro-environments. Sensors Actuators B Chem 203:54–62. https://doi.org/10.1016/j.snb.2014.06.079

Li C, Wang Z, Wang L, Zhang C (2019) Biosensors for epigenetic biomarkers detection: A review. Biosens. Bioelectron. 144:111 695. https://doi.org/10.1016/j.bios.2019.111695

Li Z, Askim JR, Suslick KS (2019) The Optoelectronic Nose: Colorimetric and Fluorometric Sensor Arrays. Chem. Rev. 119(1):231–292. https://doi.org/10.1021/acs.chemrev.8b00226

Montón H, Medina-Sánchez M, Soler JA, Chałupniak A, Nogués C, Merkoçi A (2017) Rapid on-chip apoptosis assay on human carcinoma cells based on annexin-V/quantum dot probes. Biosens Bioelectron 94:408–414. https://doi.org/10.1016/j.bios.2017.03.034

Onishi S et al (2017) Detection of bacterial fluorescence by the combination of MEMS microfluidic chip and si photodetector toward on-chip biological sensing. ECS Trans. 80(4):157–164. https://doi.org/10.1149/08004.0157ecst

Ramon C, Temiz Y, Delamarche E (2017) Chemiluminescence generation and detection in a capillary-driven microfluidic chip. Microfluid BioMEMS Med Microsyst XV 100(61): 100–610. https://doi.org/10.1117/12.2250765

Rosa AMM, Louro AF, Martins SAM, Inácio J, Azevedo AM, Prazeres DMF (2014) Capture and detection of DNA hybrids on paper via the anchoring of antibodies with fusions of carbohydrate binding modules and ZZ-domains. Anal. Chem. 86(9):4340–4347. https://doi.org/10.1021/ac5001288

Shin YH, Barnett JZ, Gutierrez-Wing MT, Rusch KA, Choi JW (2018) A hand-held fluorescent sensor platform for selectively estimating green algae and cyanobacteria biomass. Sensors Actuators B Chem 262:938–946. https://doi.org/10.1016/j.snb.2018.02.045

Sonobe H et al (2019) A novel and innovative paper-based analytical device for assessing tear lactoferrin of dry eye patients. Ocul Surf 17(1):160–166. https://doi.org/10.1016/j.jtos.2018.11.001

Stenken JA (2009) Introduction to fluorescence sensing introduction to fluorescence sensing. J Am Chem Soc 131(30)P10791–10791. https://doi.org/10.1021/ja903152j. (By Alexander P. Demchenko (National Academy of Science of Ukraine, Kiev). Springer Science + Business Media B. V.: www.springer.com. 2009. xxvi + 586 pp. $149.00. ISBN 978–1–4020–9002–8)

Su S et al (2014) DNA-conjugated quantum dot nanoprobe for high-sensitivity fluorescent detection of DNA and micro-RNA. ACS Appl Mater Interfaces 6(2):1152–1157. https://doi.org/10.1021/am404811j

Tantama M, Martínez-François JR, Mongeon R, Yellen G (2013) Imaging energy status in live cells with a fluorescent biosensor of the intracellular ATP-to-ADP ratio. Nat Commun 4. https://doi.org/10.1038/ncomms3550

Tiwari S, Vinchurkar M, Rao VR, Garnier G (2017) Zinc oxide nanorods functionalized paper for protein preconcentration in biodiagnostics. Sci. Rep. 7:1–10. https://doi.org/10.1038/srep43905

Zhang X et al (2019) Recent advances in the construction of functionalized covalent organic frameworks and their applications to sensing. Biosens. Bioelectron. 145:111 699. https://doi.org/10.1016/j.bios.2019.111699

Zhang Z, Li J, Fu D, Chen L (2013) Magnetic molecularly imprinted microsensor for selective recognition and transport of fluorescent phycocyanin in seawater. 207890. https://doi.org/10.1039/b000000x

Zhang Y, Zuo P, Ye BC (2015) A low-cost and simple paper-based microfluidic device for simultaneous multiplex determination of different types of chemical contaminants in food. Biosens Bioelectron 68:14–19. https://doi.org/10.1016/j.bios.2014.12.042

Chapter 4
Bio-microelectromechanical Systems (BioMEMS) in Bio-sensing Applications-Luminescence Detection Strategies

Ana Sofia Cerda-Kipper and Samira Hosseini

4.1 Introduction

Luminescence is emission within the optical domain of the visible, ultraviolet, or infrared light. Unlike fluorescence that may lose the activity with exposure to light over time, luminescence continues to produce its cold emission over an extended period of time (Obodovskiy (2019)). While luminescence-based analysis could be ultra-sensitive (100–1,000 times more sensitive than colorimetric or fluorescence), they may not be suitable for the multiplexed analysis (unless integrated), due to the limited types of luminescent materials (Davies et al. 2003). Luminescent biosensors have been extensively used for the detection of proteins, DNAs, RNAs, and more importantly microRNAs (Li et al. 2019). Since luminescent sensing is capable of detecting small molecules, different transduction strategies were integrated into microfluidic platforms in order to benefit from luminescent detection. A sensor of this type commonly contains a probe for immobilization of a suitable matrix, a light source for excitation of the molecules, necessary optomechanical/optoelectrical apparatuses for manipulating the signal and a readout assembly (Nagl 2015). This type of sensor relies on the luminescent dyes that alter their optical characteristics upon interaction with the target analyte. The dyes are generally embedded within a suitable network such as sensor layers, optical fiber sensors, nano-sensor particles, or magnetic nano-sensor particles (Gärtner et al. 2015). These microfluidic-integrated sensors provide data on microenvironments and phenomena that are hidden from bulk measurements, thus, it enhances our knowledge of such systems. Such devices allow simple multiplexing of experimental processes and monitoring of parameter changes (Nagl 2015). Polymer-based luminescent probes in nanoparticular shape can be used for optical sensing within microfluidic systems. Integrated sensing matrices, on the other hand, have attracted a great deal of attention and are continually growing in their

A. S. Cerda-Kipper · S. Hosseini (✉)
School of Engineering and Sciences, Tecnologico de Monterrey, Monterrey, Mexico
e-mail: samira.hosseini@tec.mx

S. Hosseini et al., *BioMEMS*, Lecture Notes in Bioengineering,
https://doi.org/10.1007/978-981-15-6382-9_4

use in different domains especially in cell culture and organ-on-a-chip applications (Pfeiffer et al. 2017). Some of the latest examples of the microfluidics BioMEMS used for luminescence detection are summarized in Table 4.1.

4.2 Luminescence Detection Strategy

Luminescence can be defined as an emission by an atom or a molecule following the absorption of light energy and entering into an excited state (Dramićanin 2018). It is triggered by the movement of electrons between different states of energy, paired with changes in the energy. The luminescent phenomenon happens when excess of energy is released by electron radiative transitions, for example by discharge of ultraviolet, visible, and near-infrared light. Based on the used energy for excitation of the electronic transitions, luminescence is categorized into different types: photoluminescence, chemiluminescence, bioluminescence, thermoluminescence, electroluminescence, etc. (Dramićanin 2018).

Luminescent sensing is based on the application of luminescent agents that alter their optical properties when in contact with the target analyte. There is a broad spectrum of applications for luminescence-based sensors, including the biology and medicine fields (De Acha et al.2017). Luminescence detection facilitate fabrication of remote (Sun et al. 2016) and non-invasive (Pasinszki et al. 2017) measurement systems for medical application as well as devices for environmental analysis. This technique has proven to have higher sensitivity than other detection methods including the colorimetric analysis (Davies et al. 2003). Some examples of luminescent biosensors are discussed in Table 4.1.

4.3 Recent Advances of Luminescence Detection in Microfluidic BioMEMS

4.3.1 Recent Advances of Luminescence Detection in Lab-On-Chip (LOC) Devices

Various efforts were made to integrate luminescent detection method into BioMEMS devices. Lab-on-chip (LOC) devices are some of the most applied platforms for luminescent biorecognition. Gärtner et al. (2015) presented a collection of microfluidic tools for cell culture while resembling in-vivo environment, thus permitting a greater understanding and a better control over cell behavior. The study demonstrated the development of a Multi-organ-tissue-flow (MOTiF) biochip (Table 4.1) based on the previous work of the authors (Raasch et al. 2015). The microfluidic cell culture toolbox included microfluidic devices with integrated membranes for separating liquid stream, and were equipped with a supply of reagents, and a 3D

Table 4.1 Recent BioMEMS platforms for luminescence detection: Type of the platform, main components, fabrication strategy, mechanism of operation

BioMEMS platform	Main components	Fabrication strategy	Mechanisms of operation	Detected analyte	Specifics	References
MOTiF biochip	• Microfluidic cell-assay toolbox • LOC handling platform • Luminescent oxygen sensor • Membrane elements • 3D feeding section	PET membranes were placed inside the injection molds. The molds were micro-structures made from COP, COC or PS by which different pieces of microfluidic platform was built. Every chamber in the device had an inlet and an outlet to control the flow. The transitional membrane acted as the support for cells while providing apical and basal nutrient source	By the aim of a handling platform, the cell culture chip remained outside a classical incubator. The fluidic connections were directly integrated in the device to connect external valves and pumps via appropriate tubing and without a direct contact with the device	No specific analyte was detected. The device was used for improving cell cultures in BioMEMS and has potential biosensing applications	The platform different cell-based assays for simultaneous monitoring of cell metabolites, and minimizing manual steps for cell culture on chip	Gärtner et al. (2015)
All-glass microreactor	• Commercially available all-glass microfluidic reactors • Luminescent chemical sensor spots	Microreactors went through a pretreatment with TPM to produce acrylate moieties on the surface and to enhance the adhesion. The chips were loaded with the prepolymer mixtures accompanied by a photo initiator and luminescent pH or oxygen probes	Chips were located on a microscope and photoinduced electron transfer contributed to the pH sensitivity of the probe. This process has led to quenching of the fluorescence at high pH	Acetylcholine and glucose	The device has certain limitations in the chemical stability when dealing with different solvents as well as a limited photostability of the luminescent probes	Pfeiffer et al. (2017)
Ratiometric H_2O_2 biosensor	• CeO_2:Eu^{3+} nanocrystals • Y_2O_3:Tb^{3+} nanoparticles	Flame spray pyrolysis was used to produce enzyme-mimetic CeO_2 nanocrystals. The nanocrystals were then doped with Eu^{3+} ions and coated on Si and glass substrates. In a one step process, the small-sized CeO_2:Eu^{3+} nanocrystals were mixed with larger, non-responsive Y_2O_3:Tb^{3+} nanoparticles in order to create a hybrid luminescent nanoaggregate for biosensing application	Eu^{3+} ions were deposited inside the CeO_2 crystalline host matrix to generate luminescent response. The catalase-mimetic activity of CeO_2: Eu^{3+}nanocrystals caused a radical quenching in the main emission peak (λ = 590 nm) when H_2O_2 was added. An EPR spectra quantified the absorption of each species	H_2O_2	The versatility of flame nanoparticles allowed the production of multicomponent systems	Henning (2019)

(continued)

Table 4.1 (continued)

BioMEMS platform	Main components	Fabrication strategy	Mechanisms of operation	Detected analyte	Specifics	References
Bandage-like O_2 sensor	• Luminescent sensing film • OLED • OPD	A drop casting method was used to make the sensing film from PtOEP, PS and TiO_2 NPs. ITO was coated on the PI substrate by using a direct *current-magnetron sputtering*. The PEDOT:PSS and IPA were combined at a 1:1 weight ratio and were deposited on the film by using spin coating. Moreover, PCBM blend solution was spin-coated as an active layer and dried. As an ETL material, PEIE was dissolved, and then coated on surface of the active layer and annealed. Finally, Al was deposited as a cathode by thermal evaporation. The OPD and OLED were combined to form a bandage-like, wearable O_2 sensor	The transcutaneous O_2 on the skin retained the sensing film where it interacts with the specific OLED component (green luminescent polymer). This results in an energy transfer that leads to the loss of luminescence which can be seen and statistically analyzed via the OPD	$tcpO_2$	This mechanism allowed the direct, noninvasive, and transcutaneous measurement of the O_2 gas pressure in human tissues. It is a low-cost platform for the simultaneous monitoring of other bio-signals including body temperature, blood pressure and CO_2 concentration	Lim et al. (2018)

The analyte of interest, and the advantages and disadvantages of each platform are presented in this table

(3-methacryloyloxypropyl) trichlorosilane (TPM); 1-(3-methoxycarbonyl) propyl-1-phenyl C_{61} (PCBM); 2,3,7,8,12,13,17,18-octaethyl-21H,23H-porphyrin platinum (II) (PtOEP); Acetylcholine (Ach); Cyclic-olefin-copolymer (COC); Cyclic-olefin-polymer (COP); Electron paramagnetic resonance (EPR); Electron transport layer (ETL); Hole transport layer (HTL); Hydrogen peroxide (H_2O_2); Multi-organ-tissue-flow (MOTiF); Organic light-emitting diode (OLED); Organic photodiode (OPD); Photoluminescence (PL); Platinum (II) tetrakis(pentafluorophenyl)porphyrin (PtTFPP); Poly (3-hexylthiophene-2,5-diyl) (P3HT); Polyethylene-terephtalate (PET); Polyethylenimine ethoxylated solution (PEIE); Polyimide (PI); Indium tin oxide (ITO); Polystyrene (PS); Titanium dioxide nanoparticles (TiO_2 NPs); Transcutaneous O_2 pressure ($tcpO_2$)

feeding section for the cultured cells. To ensure the user-friendliness of microfluidic chips, a LOC handling piece was fabricated to allow performing automated and long-term studies outside a traditional incubator. The LOC handling platform consisted of a metal or a plastic frame to place the devices and to clamp them in the frame. Furthermore, it contained a heating element, that ensured a precise temperature control. Two different inserts allowed the fluidic linking of these microfluidic platforms. The authors combined luminescent oxygen sensors with marketed polymer-based microfluidic devices for cell culture. Moreover, an integrated oxygen monitoring was applied to study oxygen consumption of microfluidic cell culture and to investigate the oxygen diffusion into the polymer chips. The advantage of using the oxygen sensor spots was that the readout could take place by an optical fiber via a contactless and contamination-free method and from outside the chip. This luminescent sensor was incorporated into the microfluidic channels, facilitating analysis of cell culture behavior while providing control over experiments. Moreover, they discussed with a previous experiment on cell culture experiments in microfluidics, that the lack of nutrients decrease proliferation, since surface area to volume-ratio is a function of cells/volume (Becker et al. 2014), hence the authors concluded that their results evidently reveal the microfluidic devices to be promising tools for several applications including fundamental cell–cell interactions study, analyzing signaling pathways, drug development and toxicity studies, and in-vitro experiments with high comparability and transferability to in-vivo tests. Therefore, the microfluidic toolbox can influence future technologies to apply microfluidics benefits to everyday cell biology analyses (Gärtner et al. 2015).

Pfeiffer et al. described a straightforward method by mask-less photopolymerization technique for incorporating luminescent chemical sensing spots into off-the shelf microreactors for determination of pH values and dissolved oxygen (DO) into all-glass microfluidic reactors, that could be achieved in less than 2 h by the aim of a microscope and a UV-LED, as suggested in Table 4.1. Even though glass microfluidic systems demonstrate larger chemical resistance, the existing techniques are limited since the bonding requirements of glass devices are commonly not the same as those for prefabricated sensing platforms. Therefore, a limited number of reports for incorporation of luminescent sensors within all glass microfluidic chips can be found in the literature (Ehgartne 2016; Lasave et al. 2015; Ungerböck et al. 2014; Mela 2005). For that reason, a straightforward and adaptable method to integrate minor sensing structures into commercially available microreactors is of great importance. The authors described two photopolymer compositions and demonstrated the function of the sensing spots in the in-line monitoring of enzymatic reactions within aqueous media. The poly (ethyleneglycol acrylate)-based sensor spots were obtained by photopolymerization of the selected oligomers in the presence of optical probes for oxygen and pH values, respectively. For the in-line monitoring setup, the microreactors were joined to pH sensing features where the readout took place using a charge-coupled device camera. Moreover, a 10× objective was applied to develop the oxygen sensor spots. The reaction could be carefully controlled by pH or oxygen measurements of the sensors, that, in turn, avoided opposing effects on the reaction turnover and on the downstream procedures. The shift in the phase angle of the luminescence signal was a function of the shelf life of the luminescence sensor spots. This shift in the angle of luminescence was utilized as the measurement signal. For observing enzymatic responses on the microscale, this method offered great potential without needing complicated instruments except for a fluorescence microscope. The fabrication of sensor structures from several prepolymer combinations was found possible. The microsensors were analyzed and have shown good sensitivity in detection and reasonable stability. Such integration methods may influence future development of flow reactors and have an impact on miniaturized cell culture and organ and tissue on a chip fields of knowledge (Pfeiffer et al. 2017).

4.4 Alternative BioMEMS for Luminescence Detection

In biomedicine, an important biomarker for inflammation is hydrogen peroxide (H_2O_2), and its quantification is important in assays engaging enzymes that produce or consume H_2O_2 linked to a particular biomarker (Pratsinis 2017). However, the optical detection of H_2O_2 has been commonly performed via peroxidase-coupled reactions using organic dyes. Due to its weak stability and/or reproducibility, this method cannot be used in-situ in multifunctional and complicated cell cultures that aim at detecting H_2O_2 levels in real-time. Henning et al. used enzyme-mimetic CeO_2 nanocrystals that are sensitive to H_2O_2 and studied the impact of H_2O_2 on electronic and luminescent nature of these nanocrystals. The authors demonstrated the performance of the biosensors by observing and recording their responses to a wide range of H_2O_2 concentrations from *S. pneumoniae*, and consequently highlighted the potential of these biosensing platforms for real-time H_2O_2 monitoring *in-vitro* within cell culture systems (Henning 2019). The authors investigated the usefulness of flame nanoparticle synthesis for the development of a ratiometric H_2O_2 biosensor based upon a double-nozzle flame spray pyrolysis of multicomponent nanoparticles performed in one step and previously developed by Strobel (2006). For nanomanufacturing processes, double-nozzle flame synthesis produced multicomponent nanoparticle systems with careful control over the size of the products and morphology independent of each nozzle. This process allowed individual control over particle growth in the intersection of each flame (Grossmann et al. 2015). However, the authors extended its use for the production of bio-responsive materials and explored the synthesis of enzyme-mimetic luminescent and H_2O_2-responsive CeO_2: Eu^{3+} nanoparticles mixed with larger luminescent nonresponsive Y_2O_3:Tb^{3+} nanoparticles (Büchel et al. 2009). To assure that the nanoparticles were mixed in the nanoscale and did not create solid componets or mixed crystal phases, the authors adjusted the two nozzles angles at 60°, in the double-nozzle setup (Fig. 4.1). Likewise, *in-situ* flame annealing was used to enhance the structural stability since it allows the porous nanoparticle films to be dipped into liquids without restructuring or loosing particle film. The main advantage of the newly developed method over the single nozzle synthesis, was that the hybrid nanoaggregates are made in a one step process minimizing the production time to half. Importantly, all constituents were present

Fig. 4.1 a The representation of double-nozzle flame reactor used in the synthesis of the nano-mixed hybrid CeO_2: Eu^{3+}/Y_2O_3: Tb^{3+} nanoaggregates. The short flame was used to produce CeO_2:Eu^{3+} nanoparticles which are smaller in size and the long flame was used for producing larger Y_2O_3:Tb^{3+} nanoparticles (the angle 60°) (Henning 2019)

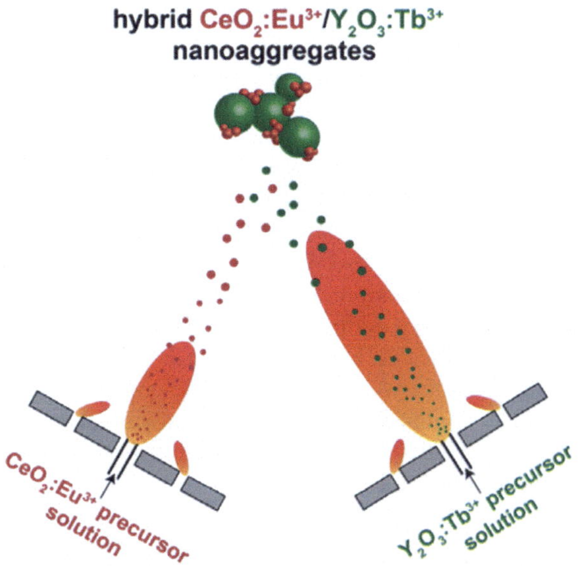

within the nanoaggregate which enabled their application as particle-based sensors in fluorescence microscopes (Henning 2019). The robust luminescence quenching when H_2O_2 was present gave the CeO_2: Eu^{3+} nanocrystals strong biosensing ability with limit of detection (LOD) value in the nM range (Pratsinis 2017) surpassing the performance of many particle-based H_2O_2 biosensors. As a result of their inorganic nature, the developed nanoaggregates exhibited high stability with respect to optical and chemical properties. The biosensor performed in realistic conditions simulated in the complex *in-vitro* bacterial cell culture platform, which could present opportunities for rapid and robust detection of H_2O_2 (Henning 2019), as suggested in Table 4.1.

Transcutaneous oxygen level is a very important parameter to diagnose and evaluate the evolution of several diseases including Raynaud disease, diabetic ulcers, and similar health conditions. Even though it has great biological and clinical relevance to study the O_2 dynamics ranging from subcellular to the macroscopic levels, few effective methods exist to non-invasively quantify O_2 in a physiological setting (Roussakis et al.2015). Lim et al. conceptualized a wearable oxygen (O_2) sensor for monitoring transcutaneous O_2 pressure ($tcpO_2$) by using luminescent gas sensing strategy integrated within wearable devices (Table 4.1). This device interacts with the oxygen present on the skin which allows an *in-vivo* constant quantitative O_2 monitoring, and likewise gives a perceptible color change via the sensing film component, for constant monitoring that facilitates the patient's treatment and recovery. The bandage-like sensor consisted of three main segments: (i) a luminescent sensing film linked to skin by using a carbon tape, (ii) an OLED light source, and (iii) an organic photodiode (OPD) light detector. The film and devices were produced by solution processes. With respect to the O_2 sensing dye and polymer matrix, 2,3,7,8,12,13,17,18-octaethyl-21H,23H-porphyrin, platinum (II) (PtOEP) and polystyrene (PS), were selected, respectively. Titanium dioxide nanoparticles (TiO_2 NPs) were integrated in PS as a light scattering center. For the working mechanism, the sensing film was placed in a close contact with the skin in order to detect $tcpO_2$ while, around the periphery of the sensor, a carbon tape was attached to the skin to optically isolate the measurement. The O_2 molecules quenched the phosphorescence in the film that resulted in the reduction of the PL intensity generated by the sensing film. The bandage-form sensor detected the dissolved O_2 which upon diffusion from the vessels to the skin (Fig. 4.2). Furthermore, the wearable O_2 sensor could screen the $tcpO_2$ in different parts of the body during exercise or work. To confirm the device performance, the authors recorded the $tcpO_2$ variants in the lower arm and a thumb by pressure-induced occlusion in the wearable sensor. The obtained data were in agreement with those measured commercially. The main beneficial features of this device were flexibility and cost-effectiveness that have rendered the sensor as a great candidate for real-time $tcpO_2$ monitoring Lim et al. 2018).

4.5 Summary

Luminescent sensing schemes are especially helpful for the recognition of small molecules and for assesing the involved parameters, and are roughly 100–1000 fold more sensitive than widely-used colorimetric strategies. The integration of these sensors into microfluidic systems allows information on microenvironments, which are very promising, and fast growing, especially in cell culture and organ-on-a-chip applications. In this chapter microfluidic devices such as lab-on-chip devices (LOC), and alternative technologies in the field of BioMEMS were reviewed. The advances in this area, opens various windows of opportunity to future developments of luminescent-based biosensors.

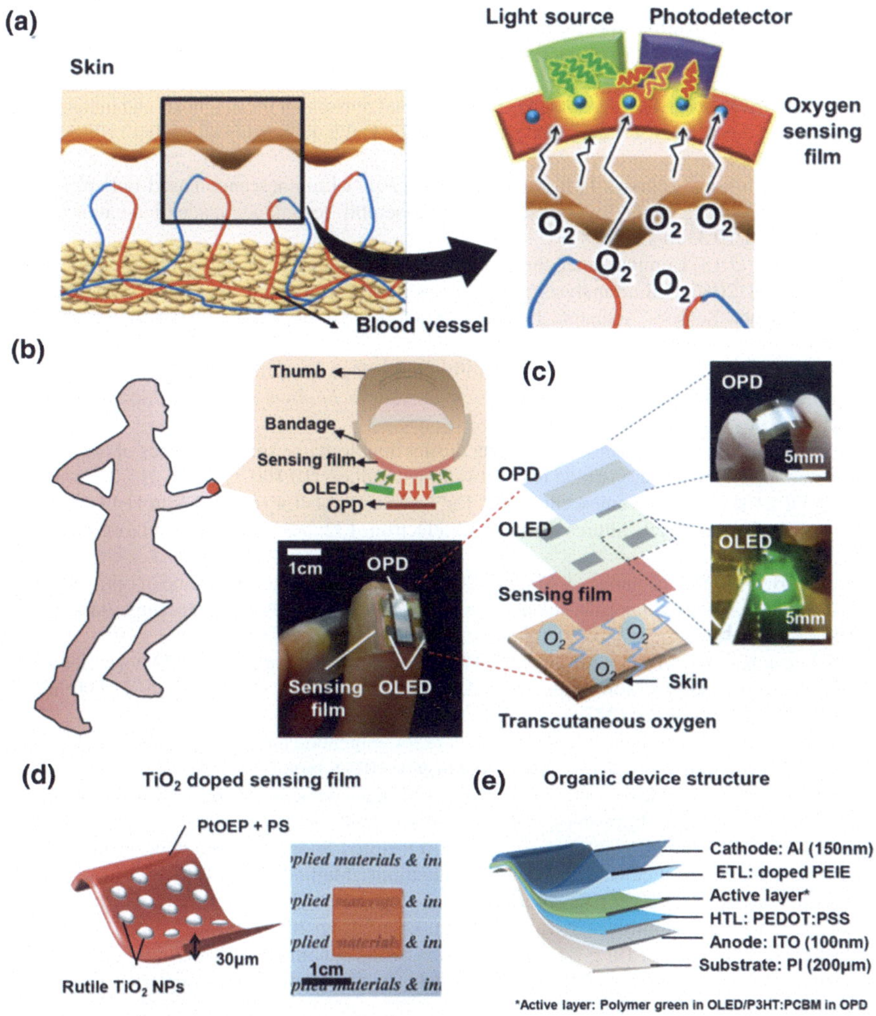

Fig. 4.2 **a** Schematics of the working mechanisms in the bandage-type O_2 sensor receiving the O_2 molecules diffused out from the blood vessels. **b** Graphic of wearable O_2 sensor with its sensing film, the OPD, and OLEDs attached to a human thumb and the digital image. **c** The structure of the platform and the design of the electrode design as well as digital photo of the flexible OPD (top) and flexible OLED (bottom). **d** Schematic of the TiO_2-doped O_2 sensing film and corresponding digital photograph. **f** The layers and structure of the flexible OPD and OLED (Lim et al. 2018)

References

Becker H, Schulz I, Mosig A, Jahn T, Gärtner C (2014) Microfluidic devices for cell culture and handling in organ-on-a-chip applications. p 89760N. https://doi.org/10.1117/12.2037237

Büchel R, Strobel R, Krumeich F, Baiker A, Pratsinis SE (2009) Influence of Pt location on BaCO3 or Al2O3 during NOx storage reduction. J Catal 261(2):201–207. https://doi.org/10.1016/JJCAT.2008.11.016

Davies R, Bartholomeusz DA, Andrade J (2003) Personal sensors for the diagnosis and management of metabolic disorders. IEEE Eng Med Biol Magaz 22(1):32–42. https://doi.org/10.1109/MEMB.2003.1191447

De Acha N, Elosua C, Matias I, Javier Arregui F (2017) Luminescence-based optical sensors fabricated by means of the layer-by-layer nano-assembly technique. academica-e.unavarra.es. https://doi.org/10.3390/s17122826

Dramićanin M (2018) Luminescence thermometry: methods, materials, and applications

Ehgartner J et al (2016) Online analysis of oxygen inside silicon-glass microreactors with integrated optical sensors. Sensors Actuators B Chem. 228:748–757. https://doi.org/10.1016/j.snb.2016.01.050

Gärtner C et al (2015) Sensor enhanced microfluidic devices for cell based assays and organs on chip. In: *Smart Biomedical and Physiological Sensor Technology XII*. https://doi.org/10.1117/12.2178690

Grossmann HK et al (2015) Nanoscale mixing during double-flame spray synthesis of heterostructured nanoparticles. J Nanoparticle Res 17(4). https://doi.org/10.1007/s11051-015-2975-8

Henning DF et al (2019) Luminescent CeO_2: Eu^{3+} nanocrystals for robust in situ H_2O_2 real-time detection in bacterial cell cultures. Biosens Bioelectron. 132:286–293. https://doi.org/10.1016/j.bios.2019.03.012

Lasave LC, Borisov SM, Ehgartner J, Mayr T (2015) Quick and simple integration of optical oxygen sensors into glass-based microfluidic devices. RSC Adv. 5(87):70808–70816. https://doi.org/10.1039/c5ra15591f

Lim CJ, Lee S, Kim JH, Kil HJ, Kim YC, Park JW (2018) Wearable, Luminescent Oxygen Sensor for Transcutaneous Oxygen Monitoring. ACS Appl Mater Interfaces 10(48):41026–41034. https://doi.org/10.1021/acsami.8b13276

Li C, Wang Z, Wang L, Zhang C (2019) Biosensors for epigenetic biomarkers detection: A review. Biosens. Bioelectron. 144:111695. https://doi.org/10.1016/j.bios.2019.111695

Mela P et al (2005) Monolayer-functionalized microfluidics devices for optical sensing of acidity. Lab Chip 5(2):163–170. https://doi.org/10.1039/b409978h

Nagl S (2015) Microfluidic platforms employing integrated fluorescent or luminescent chemical sensors: A review of methods, scope and applications. Methods Appl Fluoresc. 3(3):034003. https://doi.org/10.1088/2050-6120/3/3/034003

Obodovskiy I (2019) Luminescence. In: Radiation. Elsevier, pp 207–220

Pasinszki T, Krebsz M, Tran Tung T, Losic D (2017) Carbon nanomaterial based biosensors for non-invasive detection of cancer and disease biomarkers for clinical diagnosis. mdpi.com 17(8):1919. https://doi.org/10.3390/s17081919

Pfeiffer SA, Borisov SM, Nagl S (2017) In-line monitoring of pH and oxygen during enzymatic reactions in off-the-shelf all-glass microreactors using integrated luminescent microsensors. Microchim Acta 184(2):621–626. https://doi.org/10.1007/s00604-016-2021-2

Pratsinis A et al (2017) Enzyme-mimetic antioxidant luminescent nanoparticles for highly sensitive hydrogen peroxide biosensing. ACS Nano 11(12):12210–12218. https://doi.org/10.1021/acsnano.7b05518

Raasch M et al (2015) Microfluidically supported biochip design for culture of endothelial cell layers with improved perfusion conditions. Biofabrication 7(1). https://doi.org/10.1088/1758-5090/7/1/015013

Roussakis E, Li Z, Nichols AJ, Evans CL (2015) Oxygen-Sensing Methods in Biomedicine from the Macroscale to the Microscale. Angew Chemie Int Ed 54(29):8340–8362. https://doi.org/10.1002/anie.201410646

Strobel R, Mädler L, Piacentini M, Maciejewski M, Baiker A, Pratsinis SE (2006) Two-Nozzle Flame Synthesis of Pt/Ba/Al $_2$ O $_3$ for NO $_x$ Storage. Chem Mater 18(10):2532–2537. https://doi.org/10.1021/cm0600529

Sun C, Chen Y, Zhang G, Wang F, Liu G, Ding J (2016) Multipoint Remote Methane Measurement System Based on Spectrum Absorption and Reflective TDM. IEEE Photonics Technol Lett 28(22):2487–2490. https://doi.org/10.1109/LPT.2016.2601625

Ungerböck B, Fellinger S, Sulzer P, Abel T, Mayr T (2014) Magnetic optical sensor particles: A flexible analytical tool for microfluidic devices. Analyst 139(10):2551–2559. https://doi.org/10.1039/c4an00169a

Chapter 5
Bio-microelectromechanical Systems (BioMEMS) in Bio-sensing Applications-Bioluminescence Detection Strategies

Ana Sofia Cerda-Kipper and Samira Hosseini

5.1 Introduction

Some living organisms have the inherent property of emitting bioluminescence (BL) as a result of transforming chemical energy into light energy. It is produced by highly exothermic, enzymatically catalyzed chemical reactions where a conversion of the energy of chemical bonds to visible light takes place in organic compounds. In such reactions, molecules commonly known as luciferins (substrate) are oxidized thus generating electronically excited molecules which decay as a result of light emission. BL can also be generated by an organism itself, or by bacteria in which it cooperates in symbiosis with the host (Erzinger et al. 2017). Additionally, BL measurements do not require immediate radiation as it maintains its emission over time. In fluorescence, phototoxicity and autofluorescence can be challenging during sampling, and is commonly not advised for *in-vivo* imaging as scattering and absorption of excitation photons present a serious complication. However, BL can achieve non-invasive imaging in live samples making this technique highly favorable for various applications, such as gene regulation, gene signaling, protein–protein interactions, drug assessment, cell-based assays, molecular imaging, and non-invasive *in-vivo* imaging (Yeh and Ai 2019). When using BL in a detection system, there is no need for an outside excitation light source, since this reaction is known to have high quantum yield emission and to produce low background noise, which results in high chance of detection and sensitivity. It also represents a supreme detection strategy for miniaturized biosensing devices, by reducing weight, cost, dimension, and complexity of such integrated systems (Caputo et al. 2017). Some of the latest examples of the microfluidics BioMEMS that operate based on the BL detection strategy are provided in this chapter. A thorough comparison between these devices is provided in Table 5.1.

A. S. Cerda-Kipper · S. Hosseini (✉)
School of Engineering and Sciences, Tecnologico de Monterrey, Monterrey, Mexico
e-mail: samira.hosseini@tec.mx

S. Hosseini et al., *BioMEMS*, Lecture Notes in Bioengineering,
https://doi.org/10.1007/978-981-15-6382-9_5

Table 5.1 Recent BioMEMS platforms for bioluminescence detection: Type of the platform, main components, fabrication strategy, mechanism of operation

BioMEMS platform	Main components	Fabrication strategy	Mechanisms of operation	Detected analyte	Specifics	References
3D-Printed Microfluidic Device	• MNCs • 3D-printed helical microchannel	The 3D microfluidic device was fabricated using stereolithography. Fe_3O_4 MNCs synthesis took place by using a hydrothermal method. Capture antibodies were attached to the MNCs for oriented immobilization of the Salmonella	The free MNCs samples and MNC-EC conjugates were directed to the helical-structured microchannels. Dean drag force and lift force were applied to separate the MNC-EC from the free MNCs and signal was detected by UV–Vis absorption spectroscopy	*E-Coli*	The large size of MNC promoted more effective magnetic separation from the analyte compared to small Fe_3O_4 nanoparticles	Lee et al. (2015)

(continued)

Table 5.1 (continued)

BioMEMS platform	Main components	Fabrication strategy	Mechanisms of operation	Detected analyte	Specifics	References
LOC	• LOC system • a-Si:H diodes • ITO film	The a-Si:H diodes were coated with p-i-n structures using PE-CVD technique. The ITO film, on the other hand, was deposited by the aim of magnetron sputtering	The a-Si:H diodes acted as both temperature sensors and photosensors, while the ITO layer acted as transparent heating source. During its operation, the glass was thermally and optically coupled with the cells and electrically attached to an electronic platform, which controlled the LOC's temperature and monitored the photocurrents of the sensor	S. cerevisiae	The developed configuration allowed the on-chip detection of 1500 ± 200 genetically engineered BL yeast bioreporters cells with a linear response of 10^6 cells	Santangelo et al. (2018)

The analyte of interest, and the advantages and disadvantages of each platform are presented in this table

Amorphous silicon carbide (a-SiC:H); *E. coli* (EC); Lab-on-Chip (LOC); Hydrogenated amorphous silicon (a-Si:H); Indium tin oxide (ITO); Magnetic nanoparticle clusters (MNCs); Plasma Enhanced Chemical Vapor Deposition (PE-CVD)

5.2 Bioluminescence Detection Strategy

As discussed in Chap. 4, luminescence is the phenomenon produced by the radiation emitted by an atom or a molecule following the absorption of energy and entering into an excited state (Dramićanin 2018). It can be further divided into several categories according to the type of energy or molecule involved. Bioluminescence involves exergonic reactions of molecular oxygen with several substrates (luciferins) and enzymes (luciferases) producing photons in the visible light range (Osamu 2006). The light result varies according to the several factors including the structure of the luciferin, the amino acid sequence of the luciferase, as well as available accessory proteins. The cell biology and bioluminescence regulation vary among groups. Some examples of organisms that present bioluminescence are bacteria and fireflies. While bacteria produce light, in various other organisms the luminescence emits as flashes. To capture these flashes, a quick turn on and off is required in an enzymatic reaction (Wilson et al. 1998).

5.3 Recent Advances of Bioluminescence Detection in Microfluidic BioMEMS

5.3.1 Recent Advances of Bioluminescence Detection in Lab-On-Chip (LOC) Devices

A new method was developed for recognition of pathogenic bacteria by magnetic nanoparticle clusters (MNCs) and a helical microchannel made by 3D-printing proposed by Lee et al. The study has exhibited use of immunoassay together with the helical microchannel device (Table 5.1). For more efficient separation, the microchannel was made with a trapezoidal cross-section. Stereolithography was used to fabricate the trapezoidal cross-section and the 3D-printed device. In order to detect *E. Coli* (EC) bacteria in milk, the antibody-functionalized MNCs were used while the regular MNCs and MNC-*E. Coli* (MNC-EC) conjugates were removed from the milk by the aim of permanent magnet. The MNCs and MNC-ECs were, subsequently dispersed within a buffer solution which was inserted into a helical microchannel device regardless of the sheath flow. The MNC-ECs and MNCs were filtered and separated through the Dean drag force as well as lift force. This separation was performed in the presence of the sheath flow. After incubation of MNCs conjugated to *E. coli* antibody within the control sample, the biorecognition was performed followed by the measurement of the luminescence intensity. Since luminescence emission takes place solely in the presence of those living organisms that contain adenosine 5′-triphosphate (ATP), the luminescence intensity was a direct function of *E. Coli* concentration. Finally, the results were evaluated by UV–Vis absorption spectroscopy. An ATP luminometer was employed to confirm whether collected samples at the inner outlet were, in fact, MNC-ECs. In this specific case, the luminescence signal was detected merely from the inner outlet. The results have shown that the separation had taken place effectively. The addition of 3D-printing facilitated the fabrication of the microfluidic platform with its complicated components for successful on site detection of pathogenic bacteria Dong and Zhao (2015).

The work of Caputo et al. proposes a miniaturized lab-on-chip (LOC) favorable for monitoring living cells' activity via the on-chip BL measurement. On a single glass substrate, this system integrated hydrogenated amorphous silicon (a-Si:H) diodes which acted as temperature and light detectors, and indium tin oxide (ITO) film used as heater. Figure 5.1 shows the fully integrated thin film sensors and coupling of optical and thermal components to the sample. The proposed LOC was fabricated with a p-type doped a-SiC:H/ intrinsic a-Si:H/ n-type doped a-Si:H, a great candidate for fabrication of portable cell-based biosensors which can be used for monitoring the biological activities as well as for cell cytotoxicity assessment. The a-Si:H sensors and heaters were integrated within the same glass substrate to create a compact device for controlling the cell temperature and for sensitive detection of BL emission. The transparency of the thin film heater played a great role in transmitting the emitted light to the a-Si:H photosensors. The LOC system facilitated the thermal treatment of the cells by powering the thin ITO film, providing voltage to the film heater, and conducting heat transfer. The device also monitored the temperature through the a-Si:H temperature sensors, and enabled the BL detection by the cells. Based on these

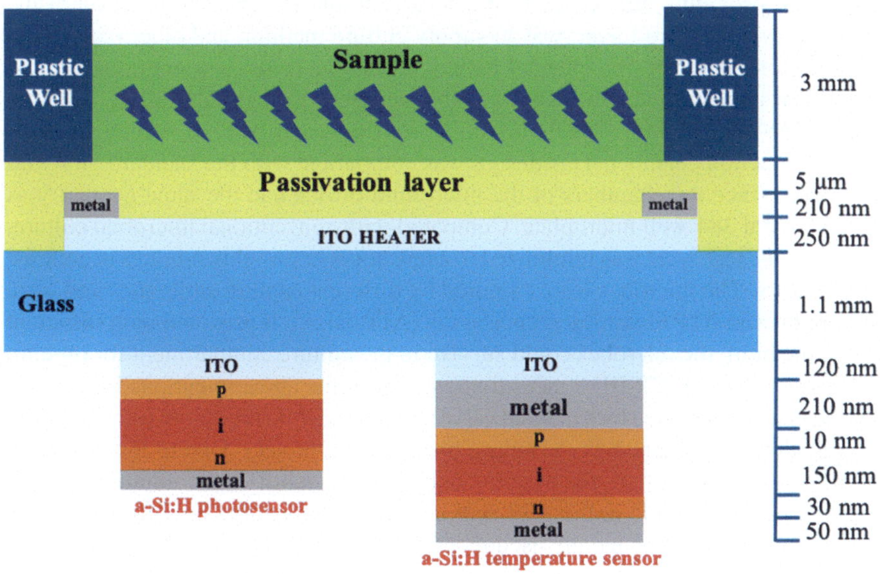

Fig. 5.1 A cross-section view of the device with the integrated layers and the two sensors (Caputo et al. 2017)

capabilities, the presented integrated LOC system could successfully monitor and control the living cells' activity (Caputo et al. 2017) (Fig. 5.1).

Dong & Zhao, developed a microfluidic simulator for personalized pathogen biorecognition by the aim of antimicrobial susceptibility testing (AST) for urinary tract infection (UTI), a commonly reported bacterial infection (Table 5.1) (Honrado and Dong 2014). The cell-based LOC system employed the immunosorbent ATP-bioluminescence assay (IATP-BLA) as a basic measuring strategy to assess the properties of uropathogenic bacteria. In this study, 13 forms of uropathogenic microbes were chosen as the target analytes (the IATP-BLA protocol is shown in Fig. 5.2). The device employed a fiberglass membrane which was sandwiched in between two polypropylene substrates in order to capture immobilized antibodies on the membrane. Due to the hydrophilicity of glass fibers, the aqueous media laterally penetrated into the neighboring reaction chambers. The analyte microbes were coupled with antibodies present on the glass fibers, and were subsequently encapsulated within a network of calcium alginate gel through the gelation reaction (Dong and Zhao 2015).

As a disposable biochip (Fig. 5.3), a microfluidic device was fabricated from two white sheets of polystyrene (PS) processed by laser ablation. A section of Waterman filter (grade GF-D fiberglass membrane) was firmly clamped between the two ablated PS layers. Different immunoglobulin Y (IgY) against specific uropathogenic microbes were immobilized on each zone on the surface of the fiberglass membrane in chambers. The upper PS component (Air Veins) was used for loading the urine sample and providing air/oxygen to the cells (Sample Layer). The function of the lower PS layer, Culture Layer, was to supply culture medium and other reagents to each reaction chamber. The channel network, Sample Veins, was engraved around the vertical through holes to lead the urine sample. Through 24 Culture Medium Veins (CMVs), the culture medium, antibiotic drugs, and ATP-BLA reagents flow into each reaction chamber. This design was compatible with the standard 384-well microplate, since the chambers of the microfluid device had the same geometry as in a traditional 384-well microplate. Compared with conventional microbial culture, the time of the test cycle was minimized to a few hours or possibly minutes in contrast to a few days. The microbes were captured by different capture antibodies and were measured via an ATP bioluminescence assay (ATP-BLA). If provided with sufficient growth medium, the microbes could survive in the culture compartment for on-chip AST former to the IATP-BLA measurement. All of the above-mentioned processes were precisely controlled by a manual valve and pump while the automation of fluidic manipulation was planned as the future phase of the project. Due to the effectiveness of on-chip IATP-BLA, the system could rapidly identify the commonly known causative agents of UTIs in the AST within 3–6 h. The authors envisioned that the medical simulator can be largely used in UTI treatment and could act as a model for the recognition and treatment of other diseases (Dong and Zhao 2015).

Santangelo et al., designed, fabricated, and tested a 3D-printed chip joined with silicon photomultipliers (SiPMs) for sensitive and simultaneous detection of luciferase BL, and for biorecognition of ATP as a model target (Fig. 5.4). The unibody-LOC (ULOC) 3D printing LOC was previously established for LOC

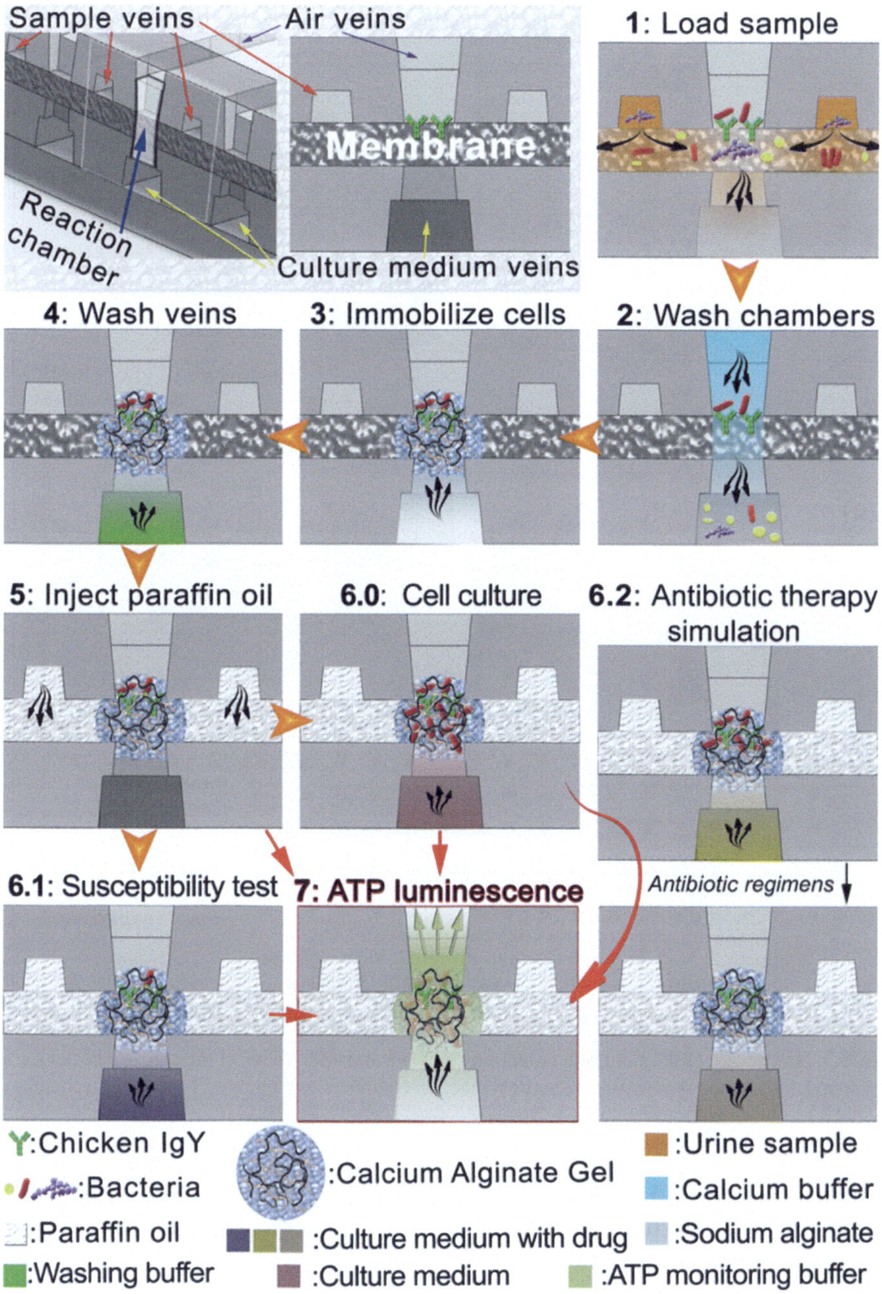

Fig. 5.2 The applied procedure of IATP-BLA test in the on-chip simulator (Dong and Zhao 2015)

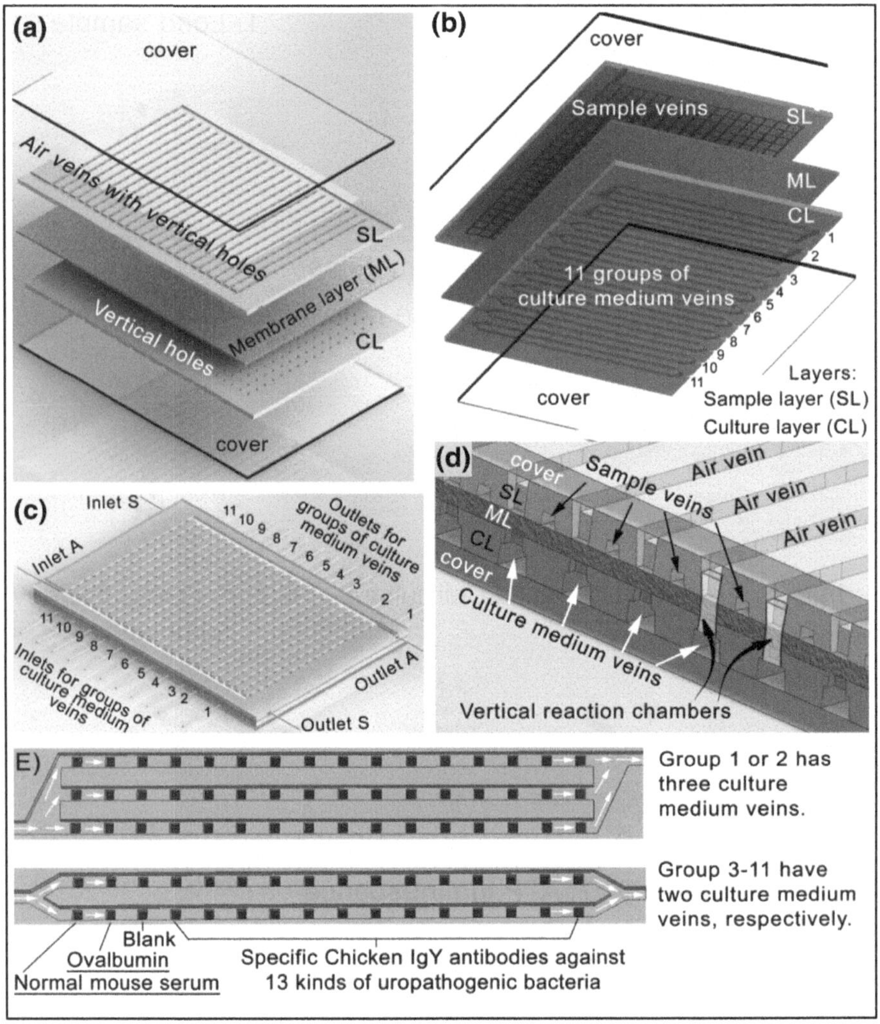

Fig. 5.3 The design of the simulator containing **a** Down and up **b** Views of the chip, **c** The device's inlets and outlets, and **d** assembly of vertical reaction chambers as well as **e** The detection channels (Dong and Zhao 2015)

(Comina et al.2014, 2015a, b), and utilized consumer grade stereolithography (SLA) 3D printers for fabrication of a monolithic printout integrating all features of the LOC functions (Table 5.1). The ports were segments of the ULOC and were developed with an outer diameter adequate to permit a tight supplement of the Si tubing. The reaction chamber and channels in the fluidic chip were open that provided an easy access for functionalization while the micro-sized surface finishing allowed sealing with conventional adhesive tape. With a strong geometric reliability even for the longer channels, the micro-channels were designed to be open on one side for the

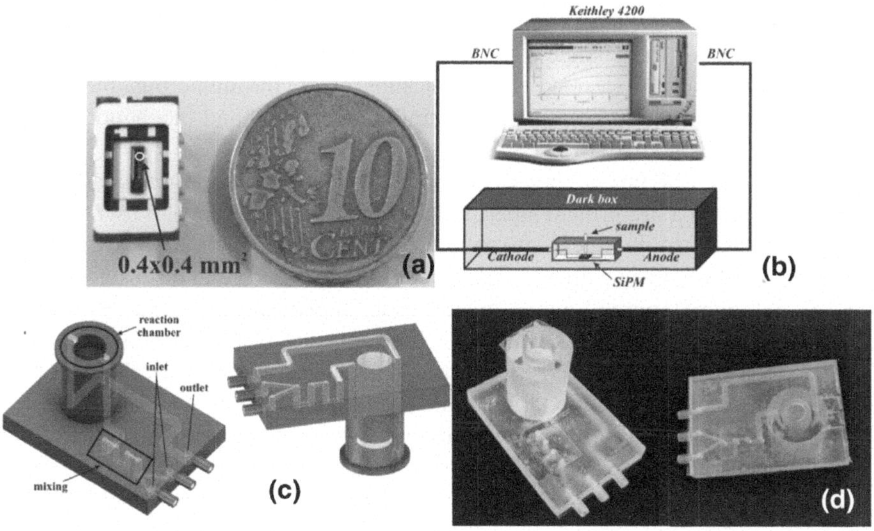

Fig. 5.4 **a** Photograph of SiPM along with **b** Schematic representation of procedure. **c** design of the 3D-printed microfluidic device along with **d** photographs of the device (Santangelo et al. 2018)

easy elimination of uncured resin. The device was attached to off-chip instrumentation via Teflon tubing which were connected to the free side of the tubing. The assay involved the pigment luciferin as the light emitting source and the enzyme luciferase to act as catalyst for luciferin oxidation. The boundaries of 3D microfluidic chip facilitated the sample delivery to nearby SiPM for enhanced BL detection efficiency. The SiPM significantly simplified the analysis compared to conventional systems yielding in a quantitative signal measurement without needing post-analysis of the images. The experiments showed that the integrated microfluidic device can be placed in close contact with environmental samples to monitor and measure biological events simultaneously and without troubling the biomolecules. The system was tested for real-time measurement and monitoring of ATP in lysate of E. coli cells and the results proved that the system's performance was not affected by the presence of lysate. Additionally, the SiPM showed great performance in monitoring ATP-BL detecting the weak BL signals produced by small ATP concentrations. The system demonstrated to be prospective for continuous-flow monitoring in different concentration levels of ATP (Santangelo et al. 2018).

5.4 Alternative BioMEMS for Bioluminescence Detection

There is an immense potential for technologies that enable environmental monitoring and medical diagnosis in under-privilaged locations. A tremendous growth in smartphone-based biosensing devices at the preliminary stage shows satisfactory sensitivity to substitute transportable light detectors including CCD and photodiodies for measurement of the analyte's medium–high concentrations. The smartphone-integrated CMOS were tested as optical biosensors for numerous biosensing methods based on fluorescence, electrochemiluminescence, and biochemiluminescence (Cevenini et al. 2016a).

Cevenini et al. (2016b) previously presented the integration of genetically engineered cells within a 3D-printed smartphone device as a low-cost and user-friendly toxicity level using bioluminescent sentinel cells (Cevenini et al. 2016b). Later, the authors reported the advancement of a BL smartphone-based cell biosensor utilizing NanoLuc luciferase reporter for measureable evaluation of anti-inflammatory activity and toxicity of grape extracts. This study used purified fusion proteins including NanoLuc as a BL donor as well as mNeonGreen fluorescent protein that acted as a receptor. The coupling of the target antibody with the analyte resulted in a color change from green–blue to blue which was received by a smartphone camera and was subsequently processed by a custom-developed app. This platform presented a reliable tool for pre-screening and sample selection for customized and effective analysis (Cevenini et al. 2016a).

5.5 Summary

Bioluminescence (BL) maintains the emission as it is a natural property in some of the living organisms and therefore offers excellent sensitivity and suitability for noninvasive in-vivo imaging. Bioluminescence represents an exceptional mechanism for the development of ultrasensitive analytical and bioanalytical protocols. The BioMEMS that operate based upon the principles of BL are commonly benefited from sensitivity, analytical speed, non-hazardous reagents, and simple procedures in a range of applications including immunoassay, protein blotting, and DNA probe assays.

References

Caputo D et al (2017) Integrated system based on thin film technologies for cell-based bioluminescence assays. In: Proceedings, vol. 1, no. 10, p. 513, Aug. 2017. https://doi.org/10.3390/proceedings1040513

Cevenini L et al (2016a) Exploiting NanoLuc luciferase for smartphone-based bioluminescence cell biosensor for (anti)-inflammatory activity and toxicity. Anal Bioanal Chem 408(30):8859–8868. https://doi.org/10.1007/s00216-016-0062-3

Cevenini L, Calabretta MM, Tarantino G, Michelini E, Roda A (2016b) Smartphone-interfaced 3D printed toxicity biosensor integrating bioluminescent 'sentinel cells.' Sensors Actuators B Chem 225:249–257. https://doi.org/10.1016/JSNB.2015.11.017

Comina G, Suska A, Filippini D (2014) Low cost lab-on-a-chip prototyping with a consumer grade 3D printer. Lab Chip 14(16):2978–2982. https://doi.org/10.1039/c4lc00394b

Comina G, Suska A, Filippini D (2015b) 3D printed unibody lab-on-a-chip: features survey and check-valves integration. Micromachines 6(4):437–451. https://doi.org/10.3390/mi6040437

Comina G, Suska A, Filippini D (2015a) Autonomous chemical sensing interface for universal cell phone readout. Angew Chemie Int Ed 54(30):8708–8712. https://doi.org/10.1002/anie.201503727

Dong T, Zhao X (2015) Rapid identification and susceptibility testing of uropathogenic microbes via immunosorbent ATP-bioluminescence assay on a microfluidic simulator for antibiotic therapy. Anal Chem 87(4):2410–2418. https://doi.org/10.1021/ac504428t

Dramićanin M (2018) Luminescence thermometry: methods, materials, and applications

Erzinger GS et al (2017) Bioluminescence systems in environmental biosensors. In Bioassays: advanced methods and applications. Elsevier pp 242–262

Honrado C, Dong T (2014) A Capacitive touch screen sensor for detection of urinary tract infections in portable biomedical devices. Sensors 14(8):13851–13862. https://doi.org/10.3390/s140813851

Lee W et al (2015) 3D-Printed micro fluidic device for the detection of pathogenic bacteria using size-based separation in helical channel with trapezoid cross-section. Sci Rep 5. https://doi.org/10.1038/srep07717

Osamu S (2006) Bioluminescence: chemical principles and methods

Santangelo MF, Libertino S, Turner APF, Filippini D, Mak WC (2018) Integrating printed microfluidics with silicon photomultipliers for miniaturised and highly sensitive ATP bioluminescence detection. Biosens Bioelectron 99:464–470. https://doi.org/10.1016/j.bios.2017.07.055

Wilson T, Hastings JW (1998) Bioluminescence CA130:120952. 14:197–230. https://doi.org/10.1146/annurev.cellbio.14.1.197

Yeh H-W, Ai H-W (2019) Development and Applications of Bioluminescent and Chemiluminescent Reporters and Biosensors. Annu Rev Anal Chem 12(1):129–150. https://doi.org/10.1146/annurev-anchem-061318-115027

Chapter 6
Bio-microelectromechanical Systems (BioMEMS) in Bio-sensing Applications-Chemiluminescence Detection Strategies

Ana Sofia Cerda-Kipper and Samira Hosseini

6.1 Introduction

An attractive detection method for the development of biosensors is chemiluminescence (CL), that identifies the presence and quantifies the concentration of an analyte by using luminescent agents that change their optical properties as a result of a chemical reaction. Over time, application of CL has become more frequent in detection of proteins and oligonucleotides in immunoassay and DNA probe assays, such as DNA sequencing and detection (Smith et al. 2018). Compared to other detection strategies including calorimetric, and fluorescence, CL offers great beneficial features thanks to its low background signal, simple readout methods, and inherent sensitivity (Xu et al. 2010). When combined with bio-microelectromechanical systems (BioMEMS), CL can offer a powerful strategy for timely and sensitive detection of various analytes in remote and rural areas and for extreme point of care (EPOC) (Wojciechowski et al. 2009).

6.2 Chemiluminescence Detection Strategy

The emission of light due to certain chemical reactions is known as CL. This emission which is commonly produced in the visible or near infrared spectral regions, is the result of an excited electronic molecular state, formed in a chemical reaction, returning to the ground state of energy through non-radiative processes (Zagatto et al. 2012; Poole 2003). Coupling of CL with other detection strategies such as fluorescent or colorimetric could offer higher sensitivity and multiply the detectable products, which, in turn, can be read via different means (Bridle 2014). In this chapter

A. S. Cerda-Kipper · S. Hosseini (✉)
School of Engineering and Sciences, Tecnologico de Monterrey, Monterrey, Mexico
e-mail: samira.hosseini@tec.mx

S. Hosseini et al., *BioMEMS*, Lecture Notes in Bioengineering,
https://doi.org/10.1007/978-981-15-6382-9_6

some of the latest biosensors advancements and strategies using CL is reviewed with a detailed comparison of these methods (Table 6.1).

6.3 Recent Advances of Chemiluminescence Detection in Paper-Based BioMEMS

The development of cheap paper-based sensors or μPADs has conferred advantageous opportunities in pursuit of bio-analytical applications that are extremely coveted in point-of-care testing (POCT) (He et al. 2011). Among paper-based techniques, Lateral Flow Immunoassay (LFIA) is one of the most common approaches in POC immunodiagnostics (Zangheri et al. 2019) due to its simplicity, robustness, portability and inexpensiveness (Qin et al. 2012). This technique is performed commonly on a nitrocellulose membrane on which specific immunoreagents are immobilized in defined positions, while sample and other reagents are transported to the detection zones in the flow driven by capillary forces. Different quantitative LFIAs were developed by instrumentally measuring the color intensity of bands or using alternative labels, such as enzymes, fluorescent nanoparticles, colorimetric markers, or electrochemical labels detected by CL (Zangheri et al. 2019). LFIAs benefit from their simplicity, high sensitivity, rapid analysis, low-power demands, and high compatibility with micromachining technologies (Li et al. 2017). In such devices, the emitted light can be imaged and analyzed by using cheap portable CCD cameras (Zangheri et al. 2019; Qin et al. 2012; Li et al. 2017).

A novel PADs biosensor for fast, sensitive, and convenient DNA detection was developed by Wang et al. (Table 6.1). A simple and rapid wax-screen-printing method was used to fabricate the device (Fig. 6.1a) followed by combining signal amplification and covalent modification. The DNA captured was immobilized, covalently on the paper zone (PADs) via the addition of N,N'-disuccinimidyl carbonate (DSC). For amplification of the detection, nano-porous gold (NPG) and carbon dots (C-dots) were used. A sandwich model in this device aimed at enhancing the wet-strength of PADs and the DNA stability on paper. Furthermore, C-dots scattered nano-porous gold (C-dots@NPG) conjugated with a DNA strand, was captured on the surface of the biosensor and employed as signal amplification label (Fig. 6.1b). The enhanced CL emission was generated in the presence of potassium permanganate and by analyzing the CL intensity the target DNA could be detected in a quantitative manner. This novel protocol allowed the combination of the PADs and CL method onto a sensitive sandwich-type CL-based DNA biosensor. The fabricated platform was low-cost, simple, portable, disposable, and easy-to-use. Under optimal conditions, this paper-based DNA sensor successfully performed with a linear range of 10^{-18} to 10^{-14} M and with a detection limit of 8.56×10^{-19} M of the target DNA. The suggested paper-based DNA sensor, with sensitive, stable, rapid, reusable and high-efficiency CL response could be a great candidate for the identification of analytes in clinical samples (Wang et al. 2013).

Table 6.1 Recent BioMEMS platforms for chemiluminescence detection: Type of the platform, main components, fabrication strategy, mechanism of operation. The analyte of interest, and the advantages and disadvantages of each platform are presented in this table

BioMEMS platform	Main components	Fabrication strategy	Mechanisms of operation	Detected analyte	Specifics	References
μPAD DNA sensor	• Microfluidic paper-based array • C-dots@NPG	The fabrication of μPADs was done via wax-screen-printing in an array of 6 columns by 3 rows. The paper was wax-screen-printed. The wax was melted and penetrated deeply into the paper to create complete hydrophobic barriers to form the micro-zones. NPG was prepared by selective dissolution of silver from silver/gold alloy via etching. The surface electrodes were modified under DSC and C-dots@NPG where used as the amplification label	Under the presence of the potassium permanganate, a sandwich-type DNA hybridization reaction occurred	DNA	Due to its simple protocol the device required no additional machinery for its operation making it a powerful potential POC testing diagnosis tool	Wang et al. (2013)

(continued)

Table 6.1 (continued)

BioMEMS platform	Main components	Fabrication strategy	Mechanisms of operation	Detected analyte	Specifics	References
CL-LFIA biosensor	• LFIA strip • Ultrasensitive cooled CCD camera • 3D printed cartridge • OFSE	Assay strips for LFIA were prepared by immobilizing, the rabbit anti-cortisol antibody (T-line) and the rabbit anti-HRP antibody (C-line), at different dilutions onto nitrocellulose membranes, which were saturated with BSA and washed with PBS including Tween 20. The membranes were subsequently assembled with adsorbent pads and sample, then cut into small segments	The oral sample was loaded via pushing the syringe plunger for filling the sample metering chamber. One of the 3 valves were opened and the HRP-cortisol conjugate and sample were transferred to the LFIA strip. The reaction mixture was sent to HRP CL substrate after incubation, and the CL signal acquisition was performed	Cortisol	The manifestation of the suitability of quantitative CL-LFIA for space applications set the stage for prospective developments of biosensing systems for multiple and early diagnosis of different analytes in space	Zangheri et al. (2019)

(continued)

Table 6.1 (continued)

BioMEMS platform	Main components	Fabrication strategy	Mechanisms of operation	Detected analyte	Specifics	References
Lab-on-Cloth	• CCGTSs • CCD camera • Plastic support	The cloth-based device fabrication was accomplished via wax screen-printing in order to create hydrophobic wax patterns	The enzyme/glucose solution resulted in oxidation of most glucose to H_2O_2. The substrate solution crossed the narrow wax barrier, and in the detection zone, combines with the enzyme/glucose solutions. The CL reaction was activated by the enzymatic oxidation solution and the emitted signals were captured by a portable CCD camera	Glucose	This device presents a low-cost, simple, easy to build and rapid microfluidic channels in cloth	Li et al. (2017)

(continued)

Table 6.1 (continued)

BioMEMS platform	Main components	Fabrication strategy	Mechanisms of operation	Detected analyte	Specifics	References
Two-channel U-shaped microfluidic device	• a-Si:H photodiodes • Al/TiW • PECVD • PDMS • PGMEA • PMMA	Thin film a-Si:H photodiode arrays were microfabricated on a glass substrate. The Al/TiW contacts for the photodiodes were deposited by magnetron sputtering and patterned with photolithography. The photodiodes were deposited by PECVD and an ITO contact layer was added on top of the photodiodes to allow electrical contact. The PDMS microfluidic platform was made through a standard soft lithography process with wet etching patterning and developed in PGMEA. This device was held on top of the photodiode chip by sandwiching them with two PMMA layers and screwing them together	OTA-BSA and OTA in the sample were in a competition to conjugate to the active binding site of the antibody. Subsequently, HRP-labeled antibody was introduced to the assay followed by luminol to produce the CL signal. This process was called icELISA	Ochratoxin A	This reusable device reduced the complication of integrated optical components as the need for external light source was omitted	Novo et al. (2013)

(continued)

Table 6.1 (continued)

BioMEMS platform	Main components	Fabrication strategy	Mechanisms of operation	Detected analyte	Specifics	References
Capillary-driven microfluidic ELISA chip	• 3D-printed dark ABS plastic • Photodiode array • Stepper motor • Si wafer	The Si wafer microfluidic chips were first dry etched by using DRIE. After priming the wafer surface with HMDS vapor, a positive photoresist was spin coated and etched on the surface. After development of the photoresist, through a "chip-olate" process, the wafer was diced and chips were separated for application	The capillary microfluidic chip was sealed with a thin layer of PDMS coated with HRP. The CL substrate was flown through the chip using capillary pressure. The light was generated in the channel once the substrate came into contact with the HRP. The signal was measured and monitored using a custom detector, which scanned the chip surface	The device has not been tested for biorecognition, however, has shown great potentials in biosensing	The device operated without a need for an external light source	Ramon et al. (2017)

(continued)

Table 6.1 (continued)

BioMEMS platform	Main components	Fabrication strategy	Mechanisms of operation	Detected analyte	Specifics	References
LOC ELISA platform	• Photodiode sensors • Photodiode chips • Microcontroller • Transimpedance amplifier • Microchannel inlets • Double-sided PCB	By using magnetron sputtering a-Si:H photodiodes arrays were microfabricated on glass substrates and patterned through photolithography and wet etched. a-SiN_x were deposited by PECVD on n-i-p a-Si:H photodiodes and patterned and etched by RIE. Subsequently, through sputtering ITO was deposited and patterned using lift-off, followed by another layer of a-SiN_x. The photodiode chips were diced and wire-bonded to the PCB. Soft-lithography was used for the development of the microfluidic device. The SU-8 substrate was prepared by immersion on PGMEA, washed using IPA and dried. PMMA were developed by laser ablation to hold the device together	By adding the target antibody through the first inlet, PBS on the second inlet, and luminol solutions on the third, the autonomous capillary ELISA was activated. To detect and quantify the CL signals, the microfluidic device was attached to a custom socket. The PMMA machined parts were used to hold and align the photodiode chips towards the microfluidic device	Anti-rabbit IgG labeled with HRP	Autonomous and sequential fluid flow capabilities such as control of the average fluid velocity at any given point of the analysis, were acquired by the capillary microfluidic device. The intricate manipulation of liquids and flow was achieved in a self-powered and automated fashion by the incorporation of simple microchannel based modules into the microfluidics device	Novo et al. (2014)

(continued)

Table 6.1 (continued)

BioMEMS platform	Main components	Fabrication strategy	Mechanisms of operation	Detected analyte	Specifics	References
Microfluidic CL immunoassay	• PMMA • Double-sided adhesive	The production of the microfluidic chips was accomplished by a laser cutting combined with hot-pressing. First, the back of the 1 mm and 2 mm PMMA board was pasted with double-sided adhesive. Laser cutting machine was used to cut the different shape of microchannels in different layers of the chip. A hot-pressing machine was used to bond the platform	The reagents were pre-loaded to the liquid storage chamber of the chip, and the vacuum suction cups were used to move the samples. The CL immunoassay was conducted and the signals were recorded and processed inside the instrument	Ferritin	The microfluidic chips were disposable, self-contained, and had a high production volume	Min et al. (2018)

(continued)

Table 6.1 (continued)

BioMEMS platform	Main components	Fabrication strategy	Mechanisms of operation	Detected analyte	Specifics	References
LOC CL immunoassay	• CCD camera • Lens • Chip • Reflector • Negative pressure sensor • Peristaltic pump • Valve actuator • Circuit board	The silicone layers were manufactured through injection molding. For the development of the tinfoil layer, PDMS was used with three parallel microchannels, and patterned with antibodies/antigens. A set of on-chip valves were fabricated with utilization of injection mold technique by using liquid PC. Finally, for the assembly of the microfluidic device, plasma treatment was employed	A solution of BSA flow through the zigzag microchannel and incubated. The CRP and mixed testosterone solution and T-mAb traveled through the zigzag microchannel to create CRP-Ab1 or T-BSA stripes on the tinfoil layer. PBST washed the zigzag microchannels and CRP-Ab2 and IgG-HRP solution were allowed into the zigzag microchannel. The last reservoir introduced the CL substrate into the zigzag microchannel, that reacted with HRP from CRP-Ab2 or IgGHRP	CRP and testosterone	The automated detection of biomarkers inside the microfluidic chips showed good reproducibility and high sensitivity	Hu et al. (2017)

(continued)

Table 6.1 (continued)

BioMEMS platform	Main components	Fabrication strategy	Mechanisms of operation	Detected analyte	Specifics	References
Tri-layer microfluidic chip	• Magnetic beads • Base-paired DNA aptamers labeled with biotins • PDMS • Glass	A CNC machined the PMMA as the master mold with the inverse microstructures. PDMS was used via replicate molding process. This process was done to create both thin and thick layers. Finally, the two layers and the glass substrate were bonded using oxygen plasma treatment	The magnetic beads were coated with the aptamers, transported to the central chamber of the device, and incubated with the blood samples. The target antibody was added and the sandwich structure was then collected by a magnetic field to wash non-binding agents. Finally, the target-aptamer-bead complex was tranferred to the NaOH chamber where the CL signal was detected	HbA1c	Automated detection during 25 min on a single chip was effectively demonstrated	Chang et al. (2015)
Smartphone-based biosensor	• LG G4 smartphone • Dark PMMA box	The dark box was constructed using black PMMA sheets, then joined using epoxy glue	To perform CL reaction, potassium permanganate reagent was added on the analyte on the TLC plate, and the emitted light was captured by a smartphone camera	Morphine	The method was simple and cost-effective	Shahvar et al. (2018)
Fiber-optic biosensor	Optical fiber	Optic fibers were cleaned and treated with piranha solution to produce hydroxyl groups	The target molecules were immobilized on the hydroxyl groups of the fibers to bind to complementary labeled detection antibodies and to record the signal	CCHF virus IgG antibodies	The protocol showed a 100-fold greater sensitivity for detection of CCHF antibodies	Algaar et al. (2015)

(continued)

Table 6.1 (continued)

BioMEMS platform	Main components	Fabrication strategy	Mechanisms of operation	Detected analyte	Specifics	References
Magnetic particles	LSA-MPs CMG-MPs	LSA-MPs, explicitly CMG-MPs, were developed by the conjugation of CMG and aminated MPs by amino/carboxyl interaction. Probes were attached to CMG-MP surfaces by covalent binding. The biotinylated amplicons were collected with MP-CMG-probes through a hybridization reaction	LSA-MP magnetically captured the target molecules, SA-AP conjugated with biotinylated amplicons. Subsequently, CL tags were released with ultrasonic treatment or DNase I treatment. These tags mixed with AMPPD were used to prolong the signal	Hepatitis B virus	Ultra-sensitivity, high selectivity, excellent reproducibility were the main features of this strategy	Yang et al. (2015)

Acrylonitrile Butadiene Styrene (ABS); Aluminum (Al); Aluminum/titanium tungsten (Al/TiW); Amorphous silicon nitride (a-SiN_X); Blood glycated hemoglobin (HbA1c); Bovine serum albumin (BSA); C-reactive protein (CRP); Carbon dots (C-dots) dotted nanoporous gold (C-dots@NPG); Carbon quantum dots (CQDs)-gold nanoparticles (AuNPs); Carboxymethylated β-1,3-glucan (CMG); Chemiluminescence (CL); CL cloth-based glucose test sensors (CCGTSs); Computer-numerical-controlled (CNC); Crimean-Congo hemorrhagic fever (CCHF); Disposable oral fluid sampling equipment (OFSE); Gold micro disk array electrodes (GD)Gravity and capillary force (GCF); Horseradish oxidase (HRP); Hydrogenated amorphous silicon (a-Si:H); Indirect competitive ELISA (icELISA); Indium tin oxide (ITO); Isopropanol (IPA); Lab-on-a-chip (LOC); Lateral flow Immunoassay (LFIA); Long spacer arm-functionalized magnetic particles (LSA-MPs); N,N'-Disuccinimidyl carbonate (DSC); ochratoxin A (OTA); Phosphate buffer solution (PBS); Plasma-enhanced chemical vapor deposition (PECVD); Polycarbonate (PC); Polydimethylsiloxane (PDMS);Polymethylmethacrylate (PMMA); Printed circuit boards (PCB); Propylene glycol methyl ether acetate (PGMEA); Propylene glycol methyl ether acetate (PGMEA); Streptavidin–alkaline phosphatase (SA-AP); Thin layer chromatography (TLC); Tris–HCl buffer solution (TBS)

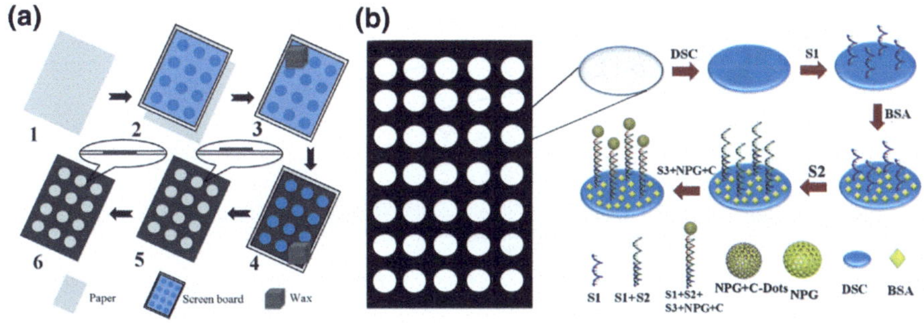

Fig. 6.1 **a** Schematic illustration of wax-screen-printing for paper microzone plate fabrication. **b** Schematic illustration of the fabrication process for the paper-based DNA sensor on paper PADs (Wang et al. 2013)

The biosensor developed by Zangheri et al. (2019) was based on the LFIA method combined with CL detection and encompassed a 3D-printed plastic cartridge comprising a sealed fluidic element with the LFIA strip (Qin et al. 2012; Li 2011; Posthuma-Trumpie et al. 2009). In this device, via pressing buttons on the cartridge, the sample flow and reagents were activated and sustained through the use of capillary forces. As a result of a collaborative project with NASA, called "IN SITU Bioanalysis" (404|NASA), the cortisol measurement in oral fluid was targeted. The astronauts are at risk of exposure to unnatural space conditions, that could increase the susceptibility to infectious diseases (Mehta et al. 2014; Taylor 2015). The biosensor was intended to fulfill the safety standards and requirements and international space station (ISS) onboard operability demanded by NASA. Assay strips for LFIA were prepared by immobilization of the antibodies on nitrocellulose membranes, which were assembled with sample and adsorbent pads (Table 6.1). The CL-LFIA assay utilized a direct competitive format (Fig. 6.2). Horseradish oxidase (HRP)-cortisol conjugate and sample loaded in the sample pad travelled along the nitrocellulose membrane via capillary force. The cortisol present in the sample rivaled with HRP-cortisol conjugate in order to bind to a limited quantity of anti-cortisol antibody immobilized on the T-line (rabbit anti-cortisol antibody). Subsequently, the anti-HRP antibody immobilized on the C-line (rabbit anti-HRP antibody) captured the excess of HRP-cortisol conjugate. Ultimately, a CL substrate (luminol/enhancer/oxidant) for HRP was added to the strip and an ultrasensitive cooled CCD camera was used to capture the image of the CL signal. In accordance with the competitive assay format, the CL signal intensity recorded at the T-line was inversely proportional to the amount of cortisol within the sample. The "IN SITU Bioanalysis" biosensor payload comprised of three components: the CL reader and its accessories, the disposable oral fluid sampling equipment (OFSE), and the disposable LFIA cartridge. Furthermore, the payload was designed to operate in microgravity and to withstand mechanical stress, such as take-off vibrations, and onboard depressurization events. The device was developed considering alterations of physical phenomena occurring in microgravity, such as bubble formation, surface wettability, and liquid evaporation. The

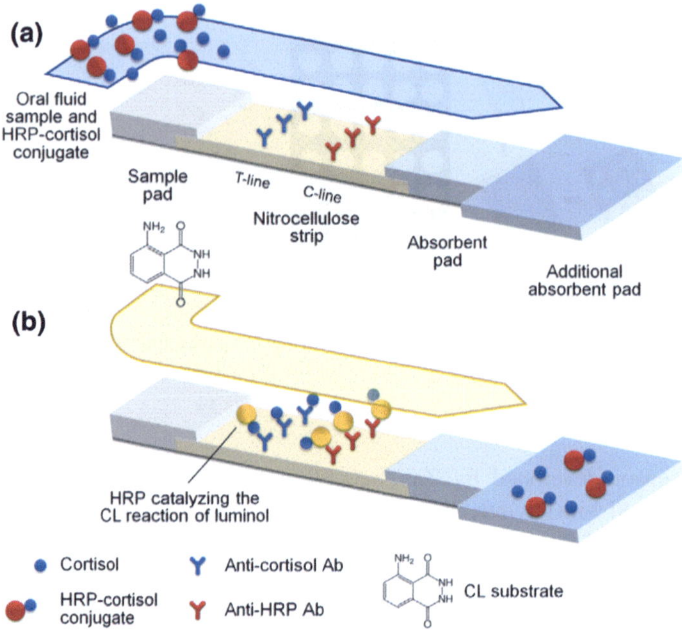

Fig. 6.2 Principle of the competitive CL LFIA assay for cortisol: **a** loading of HRP-cortisol conjugate and oral fluid sample and **b** loading of the HRP CL substrate (Zangheri et al. 2019)

disposable LFIA cartridge composed of a LFIA fluidic element is confined in a holder in order to protect the element while providing the necessary actuators which allow the astronaut to carry out the analysis through a simple manual procedure. The LFIA fluidic element was formed by a laser micromachined adhesive polypropylene layer with an engraved fluidic channels network, that was inserted between two transparent polypropylene layers. It contained the reagents required for the analysis as well as the LFIA nitrocellulose membrane, and the fluidic system needed for measuring the correct quantity of oral fluid sample while enabling the transfer of reagents and the sample on the LFIA strip. The sandwiching of polyester/acrylic adhesive layer between the polypropylene layers, benefited the sealing of the LFIA fluidic element while helping to increase the mechanical strength. The analysis followed a simple manual procedure; the flow of sample and reagents initiated via pushing a button on the cartridge and maintained by capillary forces. Astronauts could read the results directly or the data could be sent to ground personnel for processing and evaluation by medical experts. The results showed that the platform was feasible for performing sensitive detection directly onboard. This study, presented a suitable enzyme-catalyzed CL-based biosensor for ultrasensitive detection in microgravity, for the first time. The reaction was unaffected by weightless conditions, showing no further opposing effects, and the measurements at ISS appeared to be consistent with the projected cortisol levels in oral fluid. Nonetheless, the device could further be improved by assay procedure automation, and cartridge miniaturization. Moreover,

CMOS smartphone cameras can be implemented as a substitute to CCD cameras, and reusable cartridge can be replaced with disposable LFIA fluidic elements (Zangheri et al. 2019).

In the work of Li et al., a novel CL cloth-based glucose test sensor (CCGTS) was established using wax screen-printing, a simple and inexpensive fabrication method (Table 6.1) (Liu et al. 2016; Guan et al. 2015). The CL detection included enzymatic oxidation of glucose to H_2O_2 and gluconic acid and subsequent oxidizing to luminol to produce blue light in the presence of HRP. P-iodophenol (PIP) was used to heighten the CL signals which were then detected via a portable and affordable CCD camera. The layout of unfolded cloth-based device formed of a wax barrier-containing flow channel for gravity/capillary force-driven flow, a loading zone for filling the CL substrate solution, and detection zone for preloading the enzyme/glucose solutions (Fig. 6.3). The cloth device was incorporated into a well-made plastic support. Due to the substantial flexibility of the cloth device, it could be easily folded between the detection zone and the wax barrier. For the assay procedure conducted in the CCGTS, the device was placed in the CL measurement device's black box and the GOx/HRP-containing enzyme solution and the glucose solution were poured into the detection zone, respectively. Subsequently, the substrate solution comprising luminol and PIP were added onto the loading zone. The substrate solution swiftly moved in the flow channel and passed through the narrow wax barrier in the presence of capillary forces and gravity, to mix with the solutions in the detection zone. Consequently, the CL reaction was triggered, and the CL images were captured in real-time via the CCD (Fig. 6.3). The integration of wax barriers as well as gravity and capillary

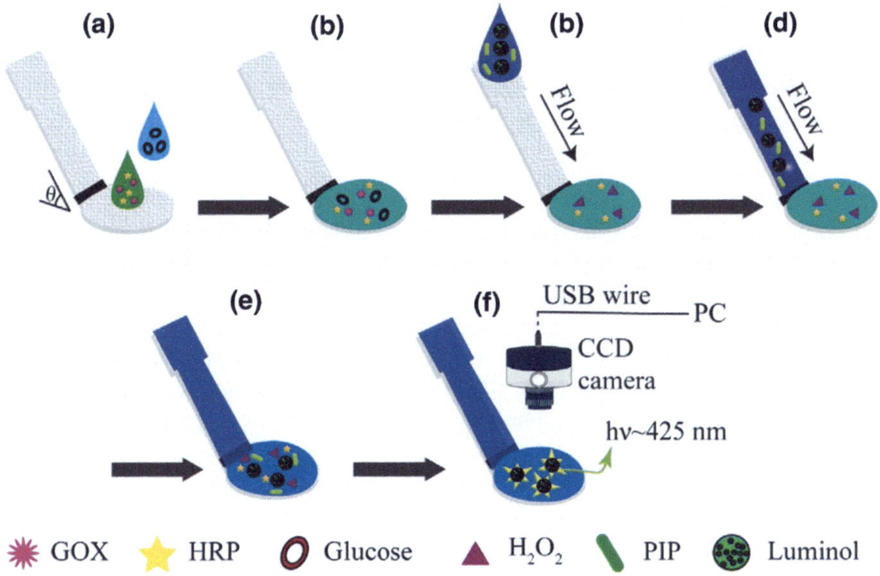

Fig. 6.3 Schematic representation of the CCGTSs for the glucose determination (Li et al. 2017)

forces in the flow channel between the detection zone and the loading zone allowed the enzyme reactions' steps to be carried out in a small-sized, single-layer cloth device. The results showed that the prepared CCGTS could offer sensitive, rapid, and quantitative biorecognition of glucose. The proposed method detected glucose over the range of 0.1–100 mM with the limit of detection (LOD) of 0.0948 mM, under optimal conditions, while the assay process was achieved in less than 5.5 min. The CCGTSs demonstrate great potentials for biochemical and medical applications, showing great promises for POCT and EPOCT (Li et al. 2017).

6.4 Recent Advances of Chemiluminescence Detection in Microfluidic BioMEMS

Over the past few decades, microfluidic devices have attracted a great deal of attention in various medical and biomedical fields including analytical chemistry, biochemistry, biodiagnosis, and POC. For miniaturized platforms, optical sensing methods are desirable, specifically, when coupled with smart devices (Nagl 2015). Such combinations of strategies simplifies the analysis process and facilitate multiplexing (Henares et al. 2008). While integration of microfluidic systems with fluorescence and colorimetry detection methods is common, CL can remarkably decrease the need of integrated optical instrumentation since it does not require any external light resources (Novo et al. 2013). Some of the latest examples of the microfluidics BioMEMS used for CL detection are as reviewed here.

6.4.1 Recent Advances of Chemiluminescence Detection in Lab-On-Chip (LOC) Devices

Novo et al. (2014) developed a novel two-channel U-shaped microfluidic for indirect competitive enzyme-linked immunosorbent assay (icELISA) system aimed at detection and quantification of ochratoxin A (OTA) (Table 6.1). A thin film of a-Si:H photodiode arrays was microfabricated on a glass substrate. The aluminum/titanium tungsten (Al/TiW) bottom contacts were deposited by magnetron sputtering and patterned by photolithography. Initially, TiW was etched by reactive ion etching (RIE) followed by Al wet etching. The n-i-p a-Si:H photodiodes were deposited by plasma-enhanced chemical vapor deposition (PECVD). Doped p-type and n-type films were developed through addition of phosphine and diborane gases to pure silane, respectively during film growth. Each of the devices were patterned and etched via RIE. The sidewalls of the device were insulated by the use of silicon nitride (a-SiN_x), deposited through PECVD. Sputtering was used to perform the deposition of a transparent top contact made of ITO enabling electrical contact to the photodiode p-layer whilst transmitting light. Lastly, a 100 nm thick passivation layer of a-SiN_x

was deposited to protect the chip. Via RIE, the contact pads were opened and the chips were diced. The chip was wire-bonded and mounted to a printed circuit board. Soft lithography was performed to fabricate the microfluidic structures.

The microchannels molds were fabricated by SU-8 on silicon. First, an aluminum physical mask was fabricated by sputtering aluminum on a quartz substrate and patterning it via wet etching. SU-8 resist was spin-coated and exposed to the UV light through the aluminum mask then baked and developed in propylene glycol methyl ether acetate (PGMEA). A second flat PDMS layer was created on top of a silicon wafer by spin coating and cured. Afterwards, the pieces were detached from the molds. A corona discharge was used to oxidize the flat 500 m thick PDMS pieces and the surface of the PDMS piece with the defined microchannels, which were then put in contact and left to seal irreversibly.

Two micromachined PMMA plates were aligned and held the microfluidic device on top of the wire bonded photodiode chip to accommodate the integration of the photodiode array and the microfluidic network. The PDMS microchannel piece was added in the top PMMA plate, comprising the microfluidic device, with the photodiodes. For the icELISA, HRP labeled anti-rabbit IgG antibodies, OTA conjugated with bovine serum albumin (BSA) and OTA were used (Fig. 6.4). The assay protocol is competitive since OTA compete with OTA-BSA for the limited anti-OTA binding

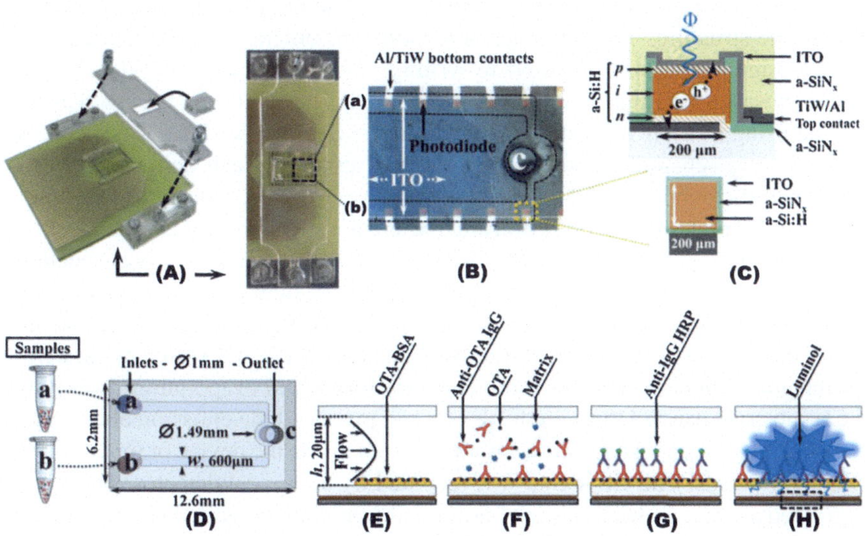

Fig. 6.4 Experimental details. **a** Incorporation of the PDMS microchannel with the photodiodes. **b** Optical micrograph indicating the alignment of the photodiode chip with the U-shaped PDMS microfluidics. **c** Top: photodiode cross section; bottom: top view schematic of photodiode. **d** The U-shape PDMS microchannel top view with inlets for reference solution (**a**) and for OTA contaminated solution (**b**); **e**–**h** schematic explanation of the microfluidic ELISA for OTA detection (Novo et al. 2013)

sites. Expectedly, a higher concentration of OTA will decrease the anti-OTA antibodies available to bind OTA-BSA molecules, hence a decrease of the icELISA. For readout, an optical microscope was used for the detection of CL light emission. The results indicated that the icELISA is a flexible and adaptable method that, when coupled with suitable sample extraction strategies, can be used for detecting OTA in complex matrices. Through this configuration, the authors successfully decreased measurement errors leading to an improvement of one order of magnitude in the LOD. The proposed platform in this study can be further developed into a future highly sensitive, portable, and fully integrated "toxin-chip" for monitoring food safety (Novo et al. 2013).

Ramon et al. (2017) developed a proof-of-concept of CL generation and detection in a capillary-driven microfluidic chip for potential immunoassay applications, as suggested in Table 6.1. Instead of using a complex assay protocol, the authors explored the CL generation under flow conditions using a simplified immunoassay model using an acridan-based reaction, catalyzed by HRP. A PDMS sealing layer was fabricated using stencil deposition and the CL substrate was flowing through the hydrophilic channels inside the PDMS device. The fabrication of the microfluidic chips involved dry etching of Si wafers using a deep reaction ion etching (DRIE) tool. Single side polished silicon wafers were used as the main substrate. After priming the wafer surface with HMDS vapor, a positive-tone photoresist was coated using a spin coater. By using a laser writing tool, the photoresist added nomenclature into the structure. This approach was more convenient for rapid fabrication of several design iterations compared to writing masks. Following the development of the exposed resist, the Si substrate was etched. The photoresist mask was then removed using a strong oxygen plasma etching for 5 min. The wafer was diced, and the chips were separated using the "chip-olate" process. Each chip comprised of loading pad surrounded with anti-wetting structures to avoid undesired spreading of a liquid placed on the pad, a detection area, where the HRP molecules were localized, and two capillary pumps that were connected in series. An air vent was connected to the last capillary pump to avoid air from escaping the flow path. Four independent channels were connected to the same loading pad and run through the detection area. Finally, the chips were sealed with the HRP coated PDMS and with a blank PDMS with integrated microbeads. By using capillary forces and evaporation-driven flow, the CL substrate was led through the device and produced the emission after coming into contact with HRP. The HRP was either coated inside the device or on the surface of the microbeads. The CL signal was recorded by an optical detector. This device has granted a better understanding of flow conditions and integrated several different strategies in a single platform (Ramon et al. 2017).

Novo et al. presented a hand-held, user-friendly POC platform, which was an incorporation of an autonomous capillary microfluidic based CL ELISA. The microfabrication of the device was performed through the use of transducers (a-Si:H photodiodes) as well as electronics implemented for data acquisition, with an integrated portable optical detection box, consisting of a microfabricated photodiode chip on glass attached to a printed circuit board (PCB) and a microcontroller (Table 6.1).

Corning display quality glass substrates were used to microfabricate the photodiodes arrays on them. Magnetron sputtering technique was used to deposit the aluminum bottom contacts that were then patterned via photolithography and wet etching technique to etch the contacts. Through PECVD, n-i-p a-Si:H photodiodes were deposited. Doped p-type and n-type films were achieved through adding phosphine and diborane, respectively. Using RIE and a mixture of SF_6 and CHF_3 gases the n-i-p stack was patterned and etched. Amorphous silicon nitride (a-SiN_x) was deposited using PECVD for insulation of the sidewalls of the diodes, and by lift-off technique, has left an access on the top of the photodiodes for electrical contact. Through sputtering method, a transparent top contact composed of ITO was deposited on the device and patterned via lift-off. In order to protect the microfabricated structures, another a-SiN_x passivation layer, was deposited via PECVD. RIE etching was performed to allow access to pads. The photodiode chips were diced and wire-bonded to a tailored designed PCBs which were positioned inside an aluminum box linked to the circuit's ground. The aluminum box consisted of two compartments: (i) a space for the microcontroller and the amplifier PCB with a port for a USB connection, (ii) and for the microfluidic devices combined with the photodiode PCB, with a lid to protect it from the electromagnetic interference as well as the external light. Soft lithography was used to fabricate the microfluidic devices and for the SU-8 mold substrate the hard mask was used. This arrangement minimized the gap between the SU-8 photoresist and the mask which led to a more accurate definition of the mold structures.

A sequence of metal sputtering, photolithography and aluminum wet etch were performed to fabricate an aluminum on glass hard mask. During the lithography, the hard mask was cut before pouring the SU-8 50 over the top of the hard mask, which was then spun, and baked. Afterwards, the SU-8 was subjected to UV and developed. Additionally, a set of PMMA plates were first fabricated via laser ablation (CO_2 laser), intended for definition of the PDMS device's bulk shape and subsequently milled in house to align the SU-8 molds with the PDMS bulk device. The SU-8 molds were then placed onto the machined PMMA plates which produced the open microfluidic inlets. The PDMS devices including the 3 inlets and the microfluidic structures and were sealed. To increase the speed of capillary pumping, a piece of absorbent paper was added to the end of the microfluidic circuit.

To demonstrate the performance of both the integrated detection system and the autonomous capillary microfluidic device, a Model IgG/anti-IgG immunoassay was employed. To perform the autonomous micro spot-based microfluidic ELISA, luminol was used for CL generation as well as the anti-rabbit IgG labeled with HRP as the target antibody (diluted in PBS). The autonomous capillary ELISA was launched through placing the target antibody, PBS, and luminol solutions at 1st, 2nd and 3rd inlets, respectively. To achieve the purpose of quantification and detection of the CL signals, the microfluidic device was implanted in the integrated setup. In addition, to align and hold the microfluidic device to the photodiode chip as an integrated LOC, a set of machined PMMA parts, with the photodiode PCB connected were utilized. The designed prototype conducted CL ELISA detection in approximately 15 min with an antibody-antigen affinity constant of 2 10^7 M^{-1} and with a LOD of 2 nM.

This prototype simplified the user interface with the device and was able to perform ELISA autonomously via sequential fluid flow due to the capillary effect. This platform exhibited great potentials for its broad range of applications in biosensing due to its flexibility of customization and the integration of detection schemes (Novo et al. 2014).

An automated microfluidic CL immunoassay platform for quantitative detection of ferritin was reported by Min et al. (2018). The material chosen for the microfluidic chip was PMMA, which is bio-friendly, cost-effective, and does not interfere with the reaction between different reagents. Laser cutting and hot press machines were used for fabrication of microfluidic chips. The single-use microfluidic chip was composed of three layers: the top layer with eight air holes, a pressure port, and an open hole; the middle layer with embedded microchannels, reagent reservoirs, a reaction reservoir, and a waste reservoir; and the bottom substrate layer. Air holes above the open reservoirs were employed to avoid reagent contamination and to load the reagents. Each reservoir was connected to an individual hydrophobic microchannel which was linked to the reaction reservoir. The microchannels were connected to the U-shaped reaction reservoir, which was in turn connected to the waste reservoir through an S-shaped pipe. The waste reservoir was filled with filter paper to ensure the flow of the waste liquid, otherwise the surface tension would hinder this waste into the left chamber. The waste chamber was linked to a pressure port.

The reliability of the automated microfluidic device was tested by measuring biomarker of ferritin by direct sandwich immunoassay. The method of acridine esterification CL was adopted to achieve the quantitative detection, and a photomultiplier tube was used to detect photons from acridine ester in alkaline conditions. The reagents were primarily pre-loaded into the liquid storage chamber of the chip. The vacuum moved the reagents across the device flexibly and easily. Subsequently, the reagents were released by the flexible vacuum suction cups that generated power by a pneumatic pump. After sample introduction, the chip was placed into a customized instrument in which the CL signals were obtained and processed. The suggested LOC platform has shown advantages including accurate quantification, sensitivity, low cost, and portability that can be promising in extreme point of care (EPOC) (Min et al. 2018).

Hu et al. reported an entirely integrated and autonomous microfluidic CL immune sensor for quantitative and automated detection of biomarkers including testosterone and C-reactive protein (CRP) in clinical samples (Table 6.1). The key components of the microfluidic device were produced by injection molding enabling low-cost mass manufacturing of the chips. The microfluidic chip was comprised of three layers: the top fluidic layer, the mid tinfoil layer that was patterned antibody/antigen stripes, and the bottom substrate layer. Each reservoir was linked to a distinct connection microchannel. The design of the channels facilitated the expansion structures to aid the operation of on-chip valves (Fig. 6.5). A through-hole was punched at the midpoint of each expansion structure, to position the on-chip valve and that 6 microchannels converged into a zigzag microchannel, which connected to a negative pressure port. Through injection molding the bottom and top layers were produced by using silicon. PDMS was utilized with three parallel microchannels to prepare

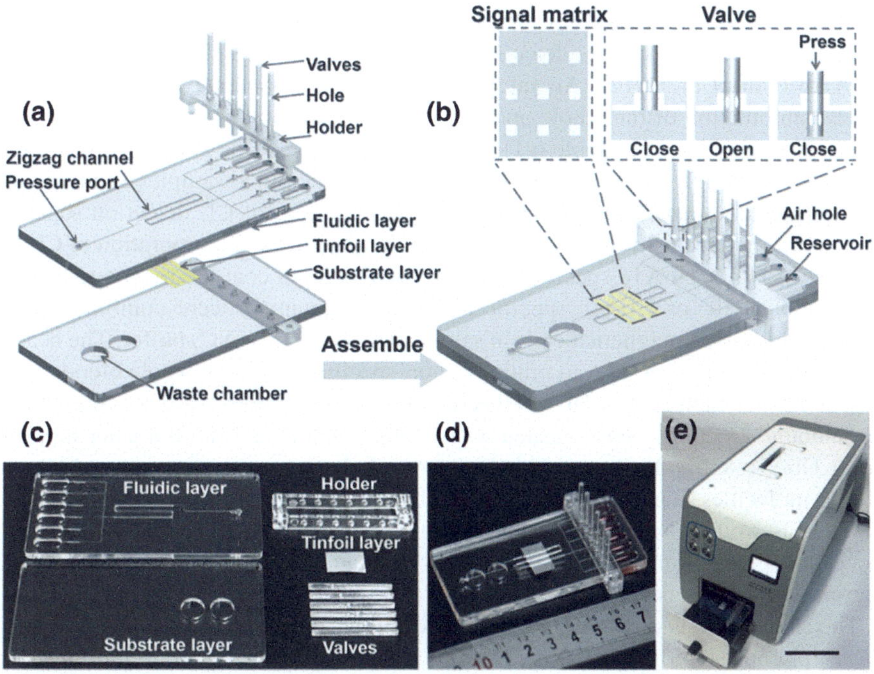

Fig. 6.5 Design layout, fabrication process and assembly of an integrated microfluidic chip equipped with on-chip valves for chemiluminescence immunoassay (Hu et al. 2017)

the tinfoil layer with patterned antibodies/antigens. Solutions of CRP-Ab1 antibody or T-BSA antigen were injected into each microchannel, incubated, and washed. Subsequently, the PDMS chip was peeled off, and to obtain the middle tinfoil layer, the tinfoil was cut with three patterned stripes. Thereupon, the set of on-chip valves including a valve holder and 6 valves were fabricated through utilization of the injection mold technique by introducing liquid polycarbonate (PC) into a mold accompanied by curing. Finally, for the assembly of the microfluidic device, plasma treatment was employed (Fig. 6.5). Before bonding the three layers, they were aligned, and the six on-chip valves were inserted and fastened. After its preparation, the chip was slid into the tailored instrument to launch the detection of testosterone or CRP (Fig. 6.5). Various reagents were pre-loaded into the respective reservoirs of the microfluidic device and segregated from the microchannels via on-chip mechanical valves where the signals were attained and processed inside the device. Ultimately, the automated microfluidic device's versatility was tested by measuring various biomarkers (testosterone by competitive immunoassay and CRP by direct sandwich immunoassay), in which the produced CL signal was picked up by means of the CCD camera within the device. The results were found highly reproducible and sensitive in the automated detection of biomarkers inside the microfluidic chips (Hu et al. 2017).

Electrochemical affinity sensors have considerable potential in developing portable analytical devices for detecting analytes such as blood glycated hemoglobin (HbA1c) (Liu et al. 2012). An integrated microfluidic system was developed to achieve automation of the whole aptamer–antibody sandwich assay to perform the measurements of HbA1c (Table 6.1) (Tankova et al. 2012). The aptamer was implemented as a main ligand to capture HbA1c or Hb in the blood samples and to facilitate the accuracy and cost reduction. The incorporation of CL detection scheme provided a higher sensitivity for the microfluidic system. For incorporation of CL to the device, Hb or HbA1c specific aptamers were used to coat the magnetic beads. Furthermore, Hb- or HbA1c specific biotinylated aptamers were combined with streptavidin-coated magnetic beads in saline-sodium citrate (SSC) buffer. The beads were then blocked with BSA to minimize non-specific binding in whole blood analysis. When the capturing of Hb or HbA1c in the blood samples were accomplished, the unbound fractions were washed away. Subsequently, acridinium ester-labeled Hb or HbA1c antibodies were added to particularly bind with the captured Hb or HbA1c molecules. Lastly, CL was produced through the addition of signal reagents of H_2O_2 as well as NaOH and a luminometer was used for its measurement. The process of aptamer–antibody sandwich assay on magnetic beads within the integrated microfluidic system is shown in Fig. 6.6a.

The chip design is shown in Fig. 6.6b. The authors developed a tri-layer microfluidic chip, consisted of two PDMS layers, comprising a thick-film and a thin-film used as an air channel layer and a liquid channel layer, respectively on top of a glass substrate. The major micro-components, such as normally-closed valves, a transport unit (a closed chamber), a waste chamber and five open chambers were integrated into this device. The microfluidic chip used a vacuum pump and an air compressor controlled via electromagnetic valves (EMVs) to automatically trigger the fluid transport for performing the assay. PDMS employed to fabricate the microfluidic chip, was created by the use of a CNC machining procedure to construct a master mold accompanied by a PDMS replica-molding process. Initially, the inverse microstructures were etched on PMMA, and after the CNC machining procedure the master mold was further polished. Lastly, to form the inverse microfluidic structures, the PDMS replica was fabricated on the PMMA master mold. For the casting process of the PDMS, the elastomer and the curing agent were combined with a weight ratio of 10:1 and a vacuum pumping process was used to remove the air bubbles. Ultimately, one glass plate and the two layers of PDMS were bonded together via utilization of an oxygen plasma treatment.

For the assay procedure (Fig. 6.6b), either Hb- or HbA1c-specific aptamers were used to pre-coat the magnetic-bead conjugates, that were then loaded into the sample chamber and combined with freeze-thawed whole blood. By manipulation of the transportation unit and the microvalve, the binding between the Hb or HbA1c in the blood sample and the aptamer took place in an incubation process for 10 min. Subsequently, an external magnet was utilized for the purpose of collecting the target-aptamer-bead complexes, whilst the nonbinding substances and the supernatant were washed away through activation of the micropump which instead directed a phosphate buffer over the waste outlet. Afterwards, to form sandwich-like structures with the

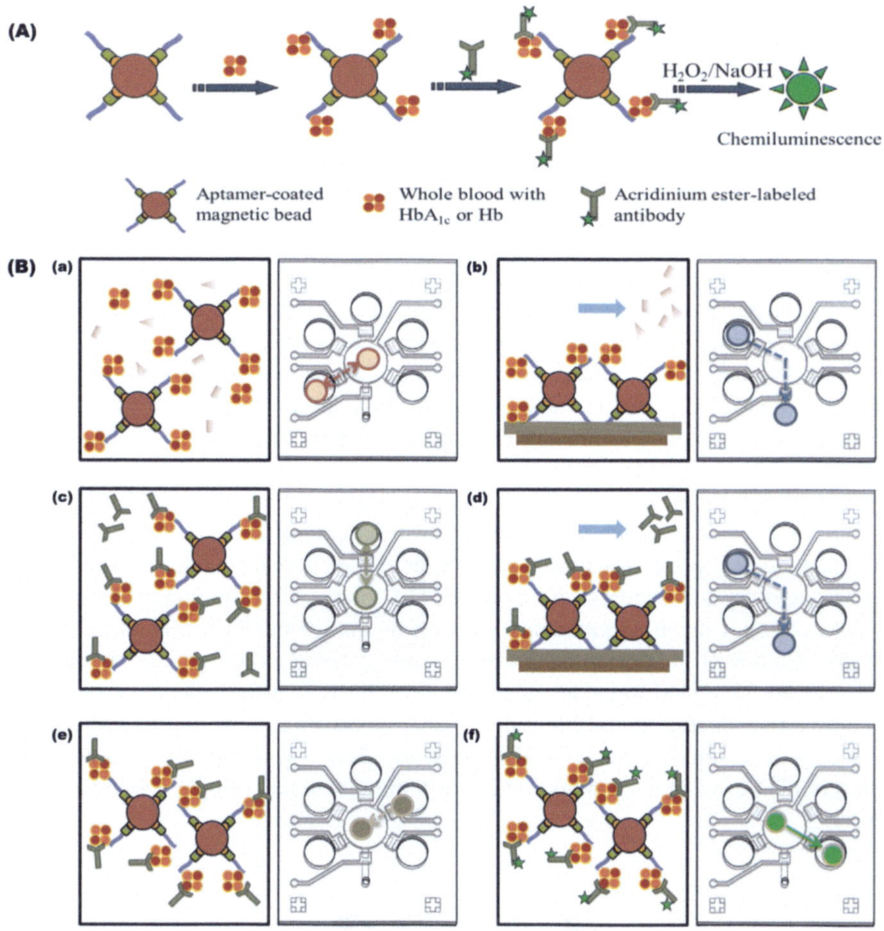

Fig. 6.6 **a** A schematic illustration of the experimental process for measuring Hb and HbA1c. **b** A schematic illustration of the experimental procedure performed on the integrated microfluidic chip (Chang et al. 2015)

target-aptamer-bead complexes in a 10-min incubation, acridinium ester-labeled Hb or HbA1c specific antibodies were injected into the transportation unit. The washing process was performed for the elimination of unbound antibodies and the magnetic collection. Finally, to re-suspend the sandwich-like complex of magnetic beads, the CL reagents were pumped into the closed chamber. Concurrently, a portable luminescence detection was implemented to measure the chemiluminescent signals while quantifying the concentrations of Hb or HbA1c. The compact microfluidic system consumed less samples and reagents and significantly shortened the detection time. The authors eliminated the use of primary antibody, which decreased the cost of the assay. The presented device has the potential to be used for diabetes screening and diagnosis at a lower cost and earlier phase to minimize the risk of diabetic complications (Chang et al. 2015).

6.5 Alternative BioMEMS for Chemiluminescence Detection

Shahvar et al. (2018), reported a low-cost, portable smartphone-based CL sensing on a thin-layer chromatography (TLC) plate for the detection of morphine. The device was fabricated using inexpensive materials and could measure the CL light's intensity radiated from the reaction between acidic potassium permanganate and morphine on a TLC plate. The authors fabricated a lab-made box (dark box) using black polymeric sheets, to prevent external light from entering the detection area and maintain a light tight setting for the CL imaging (Fig. 6.7). A movable sheet positioned in front of the box permits the simple introduction of the TLC plate into the device. A small entrance slit in front of the box was introduced using a moveable sheet, for the TLC plate to be placed at. In order to add CL reagent onto the TLC plate, a glass capillary was secured on top of the TLC plate. After the image acquisition, as the analytical signal, the intensity value of the red color (R) was chosen. Due to the dependency of the device from utilizing any external power supply, it could be used as a portable platform. The presented device was an affordable and convenient apparatus introducing an excellent alternative to existing bulky and expensive CL devices with common detectors. Moreover, the device was capable of concurrent measurement of several CL-active analytes in biological real samples (Shahvar et al. 2018).

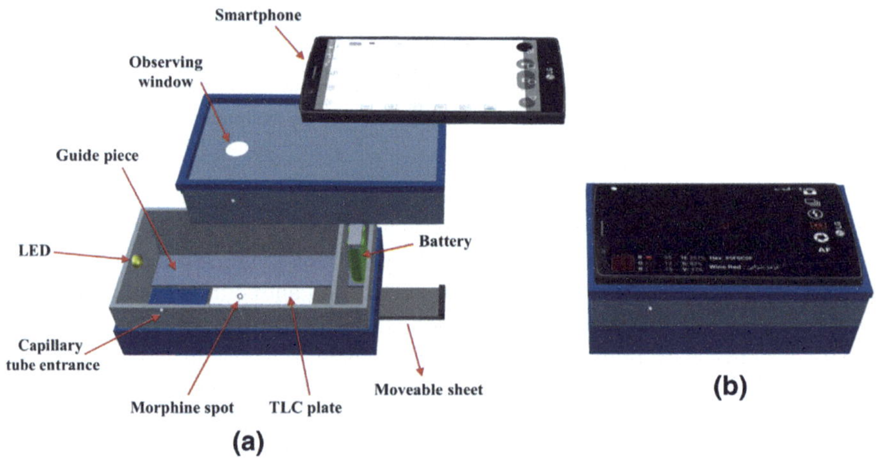

Fig. 6.7 Schematic illustration of the dark box components (**a**) and the assembled device (**b**) (Shahvar et al. 2018)

A quick and reliable diagnosis of Crimean-Congo hemorrhagic fever (CCHF) is vital for prevention of secondary spread from human-to-human, and for proper infection control measures and patient management (Papa et al. 2015; Conger et al. 2015; Pshenichnaya and Nenadskaya 2015). While diagnostics mostly rely on real-time reverse transcription polymerase chain reaction (RT-PCR) and ELISA, they require trained personnel and expensive equipment (Algaar et al. 2015; Vanhomwegen et al. 2012). Fiber-optic biosensors are a suitable alternative to existing diagnostic options for CCHF due to their small size, diagnostic accuracy, and low cost. They are quantitative and can easily be multiplexed and could be implemented on a multiple-target POC assay. Algaar et al. (2015), developed a fiber-optic biosensor for the detection of CCHF IgG antibodies (Table 6.1). They expressed a plasmid containing the nucleocapsid gene of CCHFV strain. The authors used SFS400/440B black Tefzel Superguide G UV-vis optic fibers. Tefzel jacket and silicon buffer were both removed with a fiber stripping tool, leaving a nude optical fiber core tip. The fibers were sonicated and treated with piranha solution to produce surface hydroxyl groups. Fibers' surfaces were then silanized with (3-glycidoxypropyl) trimethoxysilane. Afterward, the fibers were treated with hydrochloric acid to form vicinal diols, and with dissolved sodium m-periodate for the oxidation to aldehyde. Fibers were then rinsed with deionized water and incubated with HRP (for assay optimization) or CCHFV NP (for assay validation). Unreacted aldehyde groups were blocked using glycine and the unsaturated amines were stabilized by sodium cyanoborohydride. After exposure to a substrate, the marker enzyme oxidized it and a chemiluminescent glow was produced as a side reaction that was collected by the optical fibers and transduced to the detector. The CL detection was performed with the Immuno-star HRP chemiluminescent kit. The authors optimized the immobilization procedures to enhance the overall signal (maximize the CL output). The authors found that sonication significantly increased the biosensor endpoint signal output and that methanol washing for 20 min was the most effective for the optimization process. Moreover, the addition of 3-(10′-Phenothiazinyl) propane-1-sulfonate (SPTZ) and morpholinopyridine (MORP) increased the biosensor's output by tenfold as they provide a more durable light signal that increases the biosensor's sensitivity. The authors have enhanced the methodology by optimizing the salinization process and achieving the highest biosensor output with the temperature and duration of treatment with sodium m-peroxidate. Their experiments showed that the fiber-optic biosensor was 10-times more sensitive than colorimetric ELISA and was able to detect both patients with high and low levels of IgG antibodies which makes this platform highly desirable for early detection. The authors concluded that the assay could serve as a rapid, primary diagnostic tool for bedside CCHF diagnostics in endemic areas where extensive diagnostic equipment and trained personnel are not available (Algaar et al. 2015).

Yang et al. (2015) developed an $Fe_3O_4@SiO_2$ MP-based CL approach to detect sequence-specific DNA present in infectious pathogens, by utilizing the CL system of alkaline phosphatase (AP) and 3-(2′-spiroadamantyl)-4-methoxy-4-(3″-phosphoryloxy) phenyl-1,2-dioxetane (AMPPD) (Tang et al. 2013; Wang et al. 2012; Li et al. 2011; Yang et al. 2015) (Table 6.1). Due to the poor detection sensitivity resultant from steric hindrance caused by DNA hybridization, the authors used water-soluble carboxymethylated β-1,3-glucan (CMG) as a long spacer arm (LSA) coupled with aminated DNA-capture probes on magnetic particles (MPs). Significant enhancements were observed in DNA hybridization thanks to employment of the LSA-functionalized magnetic particles (LSA-MPs) due to the freedom that their long and flexible spacer arms provide for DNA hybridization close to the MP surface. It was observed that a limited sensitivity can be achieved due to the attenuation of CL by MPs. The authors found that, in order to allow the target to be independently detected while preventing the attenuation in CL intensity, the CL labels should be released from MPs. This shortcoming was addressed by adapting the new strategy of using covalent attachment to link the LSA-containing capture probes to MPs and resultant hybridization with target DNA fragments altered with CL labels, anticipating the release form the LSA-MPs upon LSA fracture or DNA degradation. The authors performed two methods for the release of CL labels from the LSA-MPs: DNA enzymolysis and LSA ultrasonication. The cleaned, magnetic complexes were resuspended PBS buffer and treated with ultrasonic vibration, respectively. Since DNase I nonspecifically degrades double- or single-stranded DNA to mononucleotides or oligonucleotides, Deoxyribonuclease (DNase) I was utilized for degradation of DNA to release the labels. The genomic DNA/RNA of a target pathogen was separated from the patient's serum, and through RT-PCR or PCR using biotin-11-dUTP, biotinylated amplicons were generated. Consequently, LSA-MPs were used to capture the biotinylated amplicons and combined with streptavidin–alkaline phosphatase (SA-AP). Treatment was performed with ultrasonic vibration or DNase and magnetic separation, the resultant CL labels (AP tags) converted the substrate AMPPD to AMP-D, catalytically and when exposed to the emission of a prolonged, strong CL signal, the phenoxide intermediate decomposed immediately (Fig. 6.8). With this novel method, the authors anticipated that the sensitivity of MP-based CL to be

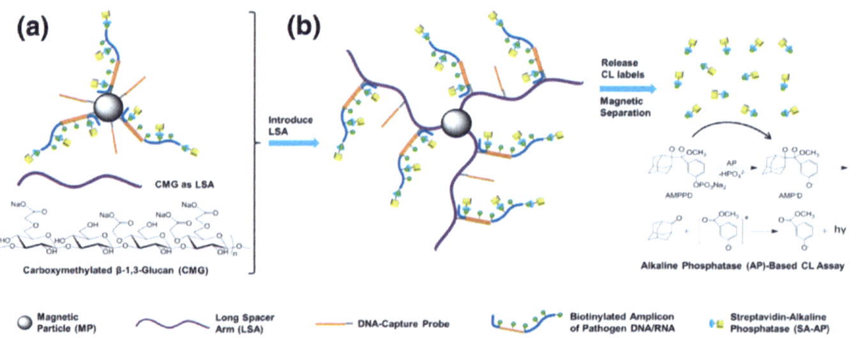

Fig. 6.8 Diagram of Improved Chemiluminescent (CL) Detection of a Target Pathogen Based on Released CL Labels from Long Spacer Arm-Functionalized Magnetic Particles (LSA-MPs). **a** MP-based CL detection method. **b** Carboxymethylated β-1,3-glucan (CMG) as LSA, forming LSA-MPs for enhanced CL detection (Yang et al. 2015)

significantly improved for the ultrasensitive detection of target pathogens in clinical samples, indicating that their method has significant promise for the early stage diagnosis of pathogenic infections (Yang et al. 2015).

Summary

Chemiluminescence (CL) has become a competitive detection method in comparison to other techniques including fluorescence and colorimetric as it is naturally generated with higher signal intensity and has higher durability for readout. By integrating CL within miniaturized BioMEMS, other beneficial features were added to such hybrid systems including compact device size, small sample volume consumption, portability, and cost-effectiveness. Enzymes are commonly incorporated or coupled with CL to amplify the detection signals or to multiply detectable products. The combination of BioMEMS and CL allows fabrication of systems that are rapid for their applications in POC or EPOC. The current chapter provides an insight on some of the latest advancements in integration of BioMEMS and CL detection method for biosensing applications.

References

Algaar F et al (2015) Fiber-optic immunosensor for detection of crimean-congo hemorrhagic fever IgG antibodies in patients. Anal Chem 87(16):8394–8398. https://doi.org/10.1021/acs.analchem.5b01728

Bridle H (2014) Optical detection technologies for waterborne pathogens. In: Waterborne pathogens. Elsevier, pp 119–145

Chang KW, Li J, Yang CH, Shiesh SC, Bin Lee G (2015) An integrated microfluidic system for measurement of glycated hemoglobin Levels by using an aptamer-antibody assay on magnetic beads. Biosens Bioelectron 68: 397–403. doi: https://doi.org/10.1016/j.bios.2015.01.027

Conger NG et al (2015) Health care response to CCHF in US soldier and nosocomial transmission to health care providers, Germany, 2009. Emerg Infect Dis 21(1):23–31. https://doi.org/10.3201/eid2101.141413

Guan W, Zhang C, Liu F, Liu M (2015) Chemiluminescence detection for microfluidic cloth-based analytical devices (μCADs). Biosens Bioelectron 72:114–120. https://doi.org/10.1016/J.BIOS.2015.04.064

He Y, Liu D, He X, Cui H (2011) One-pot synthesis of luminol functionalized silver nanoparticles with chemiluminescence activity for ultrasensitive DNA sensing. Chem Commun 47(38):10692–10694. https://doi.org/10.1039/c1cc14389a

Henares TG, Mizutani F, Hisamoto H (2008) Current development in microfluidic immunosensing chip. Anal Chim Acta 611(1):17–30. https://doi.org/10.1016/J.ACA.2008.01.064

Hu B et al (2017) Lab on a Chip An automated and portable microfluidic chemiluminescence immunoassay for quantitative detection of biomarkers. 17: 2225. doi: https://doi.org/10.1039/c7lc00249a

Li Z et al (2011) Polymerase chain reaction coupling with magnetic nanoparticles-based biotin-avidin system for amplification of chemiluminescent detection signals of nucleic acid. J Nanosci Nanotechnol 11(2):1074–1078. https://doi.org/10.1166/jnn.2011.3061

Li H, Liu C, Wang D, Zhang C (2017) Chemiluminescence cloth-based glucose test sensors (CCGTSs): a new class of chemiluminescence glucose sensors ARTICLEINFO. Biosens Bioelectron 91:268–275. https://doi.org/10.1016/j.bios.2016.12.004

Li C et al (2011) Paper based point-of-care testing disc for multiplex whole cell bacteria analysis. Biosens Bioelectron 26(11):4342–4348. https://doi.org/10.1016/J.BIOS.2011.04.035

Liu M, Liu R, Wang D, Liu C, Zhang C (2016) A low-cost, ultraflexible cloth-based microfluidic device for wireless electrochemiluminescence application. Lab Chip 16(15):2860–2870. https://doi.org/10.1039/c6lc00289g

Liu G, Khor SM, Iyengar SG, Gooding JJ (2012) Development of an electrochemical immunosensor for the detection of HbA1c in serum. Analyst 137(4):829–832. https://doi.org/10.1039/c2an16034j

Mehta SK, Laudenslager ML, Stowe RP, Crucian BE, Sams CF, Pierson DL (2014) Multiple latent viruses reactivate in astronauts during space shuttle missions. Brain Behav Immun 41:210–217. https://doi.org/10.1016/J.BBI.2014.05.014

Min X et al (2018) An automated microfluidic chemiluminescence immunoassay platform for quantitative detection of biomarkers. doi: https://doi.org/10.1007/s10544-018-0331-3

Nagl S (2015) Microfluidic platforms employing integrated fluorescent or luminescent chemical sensors: a review of methods, scope and applications. Methods Appl Fluoresc 3(3):034003. https://doi.org/10.1088/2050-6120/3/3/034003

Novo P, Moulas G, Prazeres DMF, Chu V, Conde JP (2013) Detection of ochratoxin A in wine and beer by chemiluminescence-based ELISA in microfluidics with integrated photodiodes. Sens Actuat B Chem 176:232–240. https://doi.org/10.1016/j.snb.2012.10.038

Novo P, Chu V, Conde JP (2014) Integrated optical detection of autonomous capillary microfluidic immunoassays:a hand-held point-of-care prototype. Biosens Bioelectron 57:284–291. https://doi.org/10.1016/j.bios.2014.02.009

Papa A, Mirazimi A, Köksal I, Estrada-Pena A, Feldmann H (2015) Recent advances in research on Crimean-Congo hemorrhagic fever. J Clin Virol 64:137–143. https://doi.org/10.1016/J.JCV.2014.08.029

Poole CF (2003) Instrumental aspects of liquid chromatography. In: The essence of chromatography. Elsevier, pp 431–497

Posthuma-Trumpie GA, Korf J, Van Amerongen A (2009) Lateral flow (immuno)assay: Its strengths, weaknesses, opportunities and threats. A literature survey. Anal Bioanal Chem 393(2):569–582. https://doi.org/10.1007/s00216-008-2287-2

Pshenichnaya NY, Nenadskaya SA (2015) Probable Crimean-Congo hemorrhagic fever virus transmission occurred after aerosol-generating medical procedures in Russia: nosocomial cluster. Int J Infect Dis 33:120–122. https://doi.org/10.1016/J.IJID.2014.12.047

Qin Z, Chan WCW, Boulware DR, Akkin T, Butler EK, Bischof JC (2012) Significantly improved analytical sensitivity of lateral flow immunoassays by using thermal contrast. Angew Chemie Int Ed 51(18):4358–4361. https://doi.org/10.1002/anie.201200997

Ramon C, Temiz Y, Delamarche E (2017) Chemiluminescence generation and detection in a capillary-driven microfluidic chip. In: Microfluidics, BioMEMS, and medical microsystems XV, vol 10061, p 1006100. doi: https://doi.org/10.1117/12.2250765

Shahvar A, Saraji M, Shamsaei D (2018) Smartphone-based chemiluminescence sensing for TLC imaging. Sens Actuat B Chem 255:891–894. https://doi.org/10.1016/j.snb.2017.08.144

Smith ZM, Adcock JL, Barnett NW, Francis PS (2018) Chemiluminescence—liquid phase. In: Reference module in chemistry, molecular sciences and chemical engineering. Elsevier, Amsterdam

Tang Y et al (2013) Highly sensitive and rapid detection of pseudomonas aeruginosa based on magnetic enrichment and magnetic separation. Theranostics 3(2):85–92. https://doi.org/10.7150/thno.5588

Tankova T, Chakarova N, Dakovska L, Atanassova I (2012) Assessment of HbA1c as a diagnostic tool in diabetes and prediabetes. Acta Diabetol 49(5):371–378. https://doi.org/10.1007/s00592-011-0334-5

Taylor PW (2015) Impact of space flight on bacterial virulence and antibiotic susceptibility. In: Infection and drug resistance, vol 8. Dove Medical Press Ltd., pp 249–262. doi: https://doi.org/10.2147/IDR.S67275

Vanhomwegen J et al (2012) Diagnostic assays for crimean-congo hemorrhagic fever. Emerg Infect Dis 18(12):1958–1965. https://doi.org/10.3201/eid1812.120710

Wang Y et al (2013) Facile and sensitive paper-based chemiluminescence DNA biosensor using carbon dots dotted nanoporous gold signal amplification label. doi: https://doi.org/10.1039/c2ay26485d

Wang F, Ma C, Zeng X, Li C, Deng Y, He N (2012) Chemiluminescence molecular detection of sequence-specific HBV-DNA using magnetic nanoparticles. J Biomed Nanotechnol 8(5):786–790. https://doi.org/10.1166/jbn.2012.1446

Wojciechowski JR et al (2009) Organic photodiodes for biosensor miniaturization. Anal Chem 81(9):3455–3461. https://doi.org/10.1021/ac8027323

Xu Q, Liu J, He Z, Yang S (2010) Superquenching acridinium ester chemiluminescence by gold nanoparticles for DNA detection. Chem Commun 46(46):8800–8802. https://doi.org/10.1039/c0cc03349a

Yang H, Liang W, He N, Deng Y, Li Z (2015) Chemiluminescent labels released from long spacer arm-functionalized magnetic particles: a novel strategy for ultrasensitive and highly selective detection of pathogen infections. ACS Appl Mater Interfaces 7(1):774–781. https://doi.org/10.1021/am507203s

Zagatto EAG, Oliveira C, Townshend A, Worsfold P (2012) Flow analysis with spectrophotometric and luminometric detection. Elsevier Inc

Zangheri M et al (2019) Chemiluminescence-based biosensor for monitoring astronauts' health status during space missions: results from the International Space Station. doi: https://doi.org/10.1016/j.bios.2018.09.059

404|NASA

Chapter 7
Bio-microelectromechanical Systems (BioMEMS) in Bio-sensing Applications-Biochemiluminescence Detection Strategies

Ana Sofia Cerda-Kipper and Samira Hosseini

7.1 Introduction

Luminescence is the light produced by compounds that are catalyzed by enzymes, while chemiluminescence (CL) is when the compounds undergoes specific oxidation reactions (Davies et al. 2003). Biochemiluminescence (BL-CL) is one of the most common emissions produced by a living organism and based upon the principles of CL (Roda et al. 2004). Th coupled BL-CL enzymatic reactions is used to increase the detection sensitivity in comparison to conventional colorimetric or fluorescence methods (Roda et al. 2014).

Smartphone-based devices are known for their ability to incorporate colorimetric, fluorescence, chemiluminescence, bioluminescence, and label-free detection strategies (Shahvar et al. 2018). Smart devices are considered to be the evolutionary additions to BioMEMS devices for development of point-of-care devices (Roda et al. 2014). Smart BioMEMS play a leading role in recreational activities, healthcare delivery (e.g. perform tests outside clinical laboratories) and environmental monitoring (e.g. detection of different analytes in soil and water) due to their multifunctional capabilities, imaging, and computing power, as well as their portability, affordability, simplicity, and accessibility (Shahvar et al. 2018). A fundamental advantage of BioMEMS technology is that it has the potential to offer an "all-in-one device". The built-in functions of smartphones can be further expanded through accessories that enable the sensing of different types biomarkers and analytes (Roda et al. 2014).

A. S. Cerda-Kipper · S. Hosseini (✉)
School of Engineering and Sciences, Tecnologico de Monterrey, Monterrey, Mexico
e-mail: samira.hosseini@tec.mx

S. Hosseini et al., *BioMEMS*, Lecture Notes in Bioengineering,
https://doi.org/10.1007/978-981-15-6382-9_7

7.2 Biochemiluminescence Detection Strategy

The first BL-CL based biosensors used an analyte-specific enzyme coupled with one or more "indicating" enzymes resulting in BL or CL emission (Wild 2013). One of the main advantages of BL-CL is its high detectability of the luminescence signal through high quantum efficiency (Roda et al. 2004). However, the main limitation of BL-CL-based biosensors is the different pH requirements for the enzyme reactions (Wild 2013; Microfluidic chip for biological chemiluminescence detection and detection method thereof 2012). In this chapter, we present the latest strategies reported for the development of BioMEMS that operate based on the BL-CL principle and are aimed at specific an application: biosensing.

7.3 Recent Advances of Biochemiluminescence Detection in Microfluidics BioMEMS

7.3.1 Recent Advances of Biochemiluminescence Detection in Lab-On-Chip (LOC) Devices

A patent describing a microfluidic chip for BL-CL detection was disclosed from China, to measure components inside human single blood erythrocyte (Microfluidic chip for biological chemiluminescence detection and detection method thereof 2012). The chip comprised of an interface layer, a transparent layer, a channel, a reflecting layer and a fixed layer, which were sequentially arranged from top to bottom of the device fixed by fasteners. The layers were connected by adhesives and fixed by fasteners. The interface layer and the transparent layer were provided with a liquid inlet and an outlet hole, respectively. The interface layer was comprised of an optical fiber interface. The channel layer had a liquid inlet flow channel, a micro mixer, a detecting pool, and a waste liquid buffer pool, which have sequentially communicated with each other. The optical fiber interface was arranged above the detecting pool, while the bottom of the optical fiber interface was connected with a light penetration layer. The detecting pool was in contact with the lower surface of the channel layer and the bottom of the detecting pool was connected to the reflecting layer. Moreover, the bottom of the waste liquid buffer pool was connected to the fixed layer. The fluorescence micro-spectrum detection was performed in a single step flow passage structure, thus improving luminous intensity by enhancing solution mixability, and optical/light detection efficiency of BL-CL detection. This microfluidic chip has shown the advantages of simple fabrication method, convenient operation, rapid detection, high sensitivity, and accurate detection result, as described in Table 7.1.

Another patent from Japan used cartridge for measurement of BL-CL (Measuring method, cartridge for measurement, and measuring device 2013). The measurement took place through injecting a sample into a reaction tank via an inlet side in the flow

Table 7.1 Recent BioMEMS platforms for biochemiluminescence detection: Type of the platform, main components, fabrication strategy, mechanism of operation. The analyte of interest, and the advantages and disadvantages of each platform are presented in this table

BioMEMS platform	Main components	Fabrication strategy	Mechanisms of operation	Detected analyte	Specifics	References
Microfluidic chip	• Micro mixer • Detecting pool • Waste liquid buffer pool • Optical fiber interface	Through a proprietary method several layers of microfluidic were aligned and attached to each other in order to conduct the assay	The reaction mixture was inserted to the device via the inlet. All different layers of the device were in communication. After the assay process was followed the signal readout took place above the detection pool	Single blood erythrocyte	The device demonstrates a rapid and highly sensitive method	Microfluidic chip for biological chemiluminescence detection and detection method thereof (2012)
Cartridge-based BL-CL measurement	• Cartridge • Measurement apparatus • Reaction vessel	Through a proprietary method, a cartridge was equipped with an inlet and an outlet as well as a reaction vessel	The dry reagents were stored in different concentrations inside the cartridge. A solvent was transferred through the device dissolving the dry samples thus BL-CL reaction	Endotoxin and β-glucan	The device demonstrated a low chance of contamination of reagents. However, a major challenge was preventing the deterioration of the reagent containing biological compounds	Measuring method (2013)

(continued)

Table 7.1 (continued)

BioMEMS platform	Main components	Fabrication strategy	Mechanisms of operation	Detected analyte	Specifics	References
SmartBA and SmartChol	• iPhone 5S • BI-CMOS sensor • 8MP camera • Smartphone accessory • Minicartridge	The minicartridge and the mini dark box smartphone accessory were fabricated using a dual-extrusion 3D printer using black PMMA thermoplastic. The minicartridge was printed in two separate pieces, which were glued together	SmartBA relied on conversion of certain functional groups that emitted BL-CL in the presence of bacterial luciferase. SmartChol, operated based on coupled enzymatic reaction which generated BL-CL in the presence of luminol$-H_2O_2-$HRP. In both cases, the signal was recorded by smart phone	Cholesterol and BA	SmartBA and SmartChol could be considered as the pioneer devices in the integration BL-CL detection and smartphones for POC analysis	Roda et al. (2014)

Horseradish peroxidase (HRP); Total Bile Acid Smartphone-Based Assay (SmartBA); Total Cholesterol Smartphone-Based Assay (SmartChol)

path and communicating with the outlet. The dry reagents were stored in the device in the right portion and a solvent traveled between the inlet and the outlet inside the reaction vessel. As the solvent passed through the channel, the dry reagents were dissolved and transferred to the reaction vessel. The optical characteristics of the reaction mixture within the reaction vessel were analyzed by BL-CL measurement. This measurement method was used for endotoxin and β-glucan detection. A main concern in fabrication and reagent storage of this device was the deterioration of the reagent containing biological compounds over time. It was, therefore, recommended to use the dissolved reagents immediately after preparation.

7.4 Alternative BioMEMS for Biochemiluminescence Detection

Roda et al. (2004) for the first time, reported the use of a smartphone to image and quantify BL-CL coupled bio-specific enzymatic reactions to detect analytes in biological fluids. By using low-cost 3D printing technology, the authors fabricated a smartphone accessory and a minicartridge for hosting bio-specific reactions (Fig. 7.1). The device consisted of two main parts: a phone adapter with a dark box and lens holder and a cartridge for the bioassays, specifically designed to match an iPhone 5S. This accessory played a dual role of acting as a dark box for shedding from ambient light and hosting the minicartridge, which could be customized according to the target chemical reactions and diagnostic needs. The disposable mini-cartridge contained a blood separator pad holder, with a LF1 glass fiber filter, connected to a reaction chamber where a nitrocellulose disk supporting the specific enzymes was placed. A separate reservoir for BL/CL reagents was connected via microfluidics to the reaction chamber in order to prevent premature mixing. Two assays were conducted in this device: (i) Total Bile Acid Smartphone-Based Assay (SmartBA), a bioluminescence assay for measurement of total bile acids (BA) using 3α-hydroxyl steroid dehydrogenase co-immobilized with bacterial luciferase system and a chemiluminescence substrate; (ii) and Total Cholesterol Smartphone-Based Assay (SmartChol), for total cholesterol measurement using cholesterol esterase/cholesterol oxidase coupled with the luminol–H_2O_2–horseradish peroxidase (HRP).

In SmartBA, converted groups from BA 3αhydroxyl group, in the presence of bacterial luciferase, reacts with decanal and oxygen to produce flavin mononucleotide (FMN), decanoic acid, and light. The light intensity was proportional to BA concentration in the initial reaction. The smartphone camera was used to image and quantify the light produced by BL-CL reactions and to amplify analyte-specific enzymatic reactions. Due to the design of the device, no accidental release of the reagents could occur even when turning the minicartridge upside down.

The SmartChol assay was based on coupled enzymatic reactions in two steps. Firstly, esterification of cholesterol hydrolysis was done by cholesterol esterase and cholesterol oxidation by cholesterol oxidase. Secondly, the CL detection of

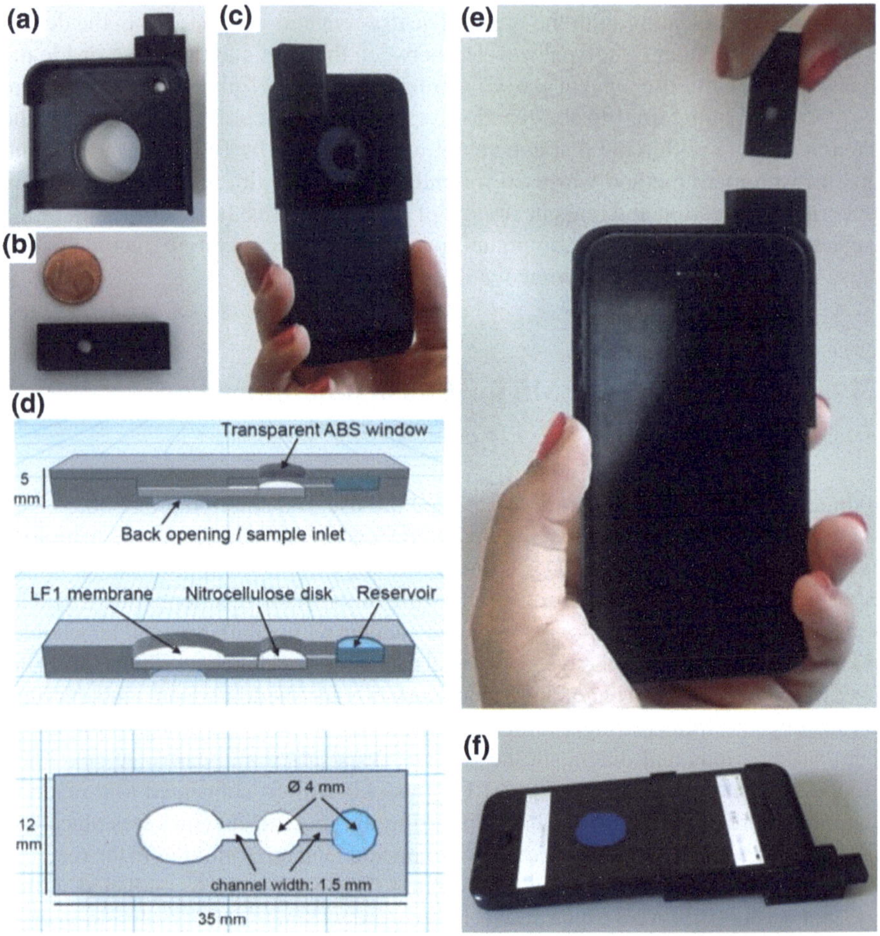

Fig. 7.1 **a** Picture of the accessory, **b** picture of the minicartridge, and **c** picture of the accessory snapped into the smartphone. **d** Schematic cutaway drawings of the minicartridge showing the integration of the various components. The transparent ABS optical window (200 μm) of 4 mm diameter allows imaging of biochemiluminescent reaction. **e** Introduction of the minicartridge into the accessory and **f** picture of a representative CL acquisition with the smartphone (Roda et al. 2014)

the produced hydrogen peroxide was perfromed by using Super Signal West Dura Luminol/ Enhancer solution in the presence of HRP as a catalyst (Roda et al. 2014). SmartBA and SmartChol could be considered as the pioneer devices in the integration BL-CL detection and smartphones for POC analysis.

7.5 Summary

Biochemiluminescence (BL-CL) is one of the most sensitive methods for detecting analytes. This detection method is suitable for a wide range of applications in the biotechnological field including reporter gene technology, gene probe assays, and immunoassays. Moreover, due to its powerful emission, it can target analytes for the medical, pharmaceutical and environmental applications and within miniaturized, micromachined, bioanalytical devices for high throughput analysis. In this chapter, some of the latest technologies developed based upon BL-CL detection technique are presented in great detail.

References

Davies R, Bartholomeusz DA, Andrade J (2003) Personal sensors for the diagnosis and management of metabolic disorders. IEEE Eng Med Biol Mag 22(1):32–42. https://doi.org/10.1109/MEMB.2003.1191447

Measuring method, cartridge for measurement, and measuring device, July 2013

Microfluidic chip for biological chemiluminescence detection and detection method thereof, May 2012

Roda A, Pasini P, Mirasoli M, Michelini E, Guardigli M (2004) Biotechnological applications of bioluminescence and chemiluminescence. Trends Biotechnol 22(6):295–303. https://doi.org/10.1016/J.TIBTECH.2004.03.011

Roda A, Michelini E, Cevenini L, Calabria D, Calabretta MM, Simoni P (2014) Integrating biochemiluminescence detection on smartphones: Mobile chemistry platform for point-of-need analysis. Anal Chem 86(15):7299–7304. https://doi.org/10.1021/ac502137s

Shahvar A, Saraji M, Shamsaei D (2018) Smartphone-based chemiluminescence sensing for TLC imaging. Sens Actuat B Chem 255:891–894. https://doi.org/10.1016/j.snb.2017.08.144

Wild D (2013) The immunoassay handbook: theory and applications of ligand binding. ELISA—Google Libros

Chapter 8
Bio-microelectromechanical Systems (BioMEMS) in Bio-sensing Applications-Electrochemiluminescence Detection Strategies

Ana Sofia Cerda-Kipper and Samira Hosseini

8.1 Introduction

The biosensors that operate based upon electrochemical detection method incorporate an electrochemical transducer, which mediates the generation of a measurable electrical signal (Li et al. 2019). Electrochemiluminescence (ECL) was used to monitor enzymatic reactions, and for analysis of various species (Richter 2008) and analytes (Ju et al. 2017), as well as in different biosensor applications (Roda 2017). ECL-based immunosensing combines the unique advantages of highly specific immunoreaction and convenient ECL biosensing offering a promising method for analyte measurement (Ju et al. 2017). ECL labels have distinct advantages over other detection methods as they are sensitive, nonhazardous, and inexpensive. These labels contribute to the diagnostic of a particular element, are linear over a wide range, and incorporate a relatively inexpensive equipment with modest operation (Richter 2008; Ju et al. 2017; Roda 2017; Zhang 2019). Additionally, ECL-based biosensors are simple, rapid, and inexpensive in comparison to other methods, such as surface-enhanced Raman spectroscopy (SERS) (Li et al. 2019). ECL detection technique has a wide range of applications in multiple fields including environmental monitoring, clinical diagnosis, food safety, biodiagnosis, and pharmacology (Zhang 2019; Farokhi-Fard et al. 2019). Moreover, due to its efficiency, considerably low detection limit, reducing the LOD, simplicity, and cost-effectiveness, ECL has become an excellent technique for development of bio-microelectromechanical systems (BioMEMS) for point-of-care (POC) or extreme point of care (EPOC) (Socorro-Leránoz et al. 2019; Blair and Corrigan 2019).

A. S. Cerda-Kipper · S. Hosseini (✉)
School of Engineering and Sciences, Tecnologico de Monterrey, Monterrey, Mexico
e-mail: samira.hosseini@tec.mx

S. Hosseini et al., *BioMEMS*, Lecture Notes in Bioengineering,
https://doi.org/10.1007/978-981-15-6382-9_8

8.2 Electrochemiluminescence Detection Strategy

The ECL signal is generated when the chemiluminescent (CL) phenomenon is triggered by an electrochemical method without a need for an external light source (Hu and Reviews 2010; Bard 1988). This detection strategy offers high versatility, a wide dynamic detection range, low background noise, simple optical setup, and good reproducibility (Ju et al. 2017). Furthermore, various reactants can be electrochemically regenerated at the electrode. The regeneration of these reactants allows them to take part in ECL reactions again in an excess of co-reactants that can react with luminophores to produce higher light intensity. As a result, many photons are produced per each measurement cycle. This greatly enhances the sensitivity of the technique and classifies it as an excellent candidate for biosensors (Hu and Reviews 2010). In this chapter, we review and compare some of the latest advancements of the BioMEMS platforms that have integrated the ECL strategy as a means for analyte recognition.

8.3 Recent Advances of Electrochemiluminescence (ECL) Detection in Paper-Based BioMEMS

Since its development, μPADs have shown tremendous prospects in the molecular analysis of biological fluids (e.g. serum, blood, and urine), health monitoring and environmental analysis in developed and developing countries, as well as in resource-limited and remote parts of the world. Particularly, μPADs integrated into analytical systems have advanced into refined applications that involve colorimetric, fluorescence, luminescence, electrochemical, chemiluminescence, and electrochemiluminescence detection strategies (Yang 2014). Conventional colorimetric approach for qualitative analysis of analytes on μPADs presents a limited sensitivity. However, ECL incorporates the gains of electrochemistry and chemiluminescence to offer a powerful method for highly sensitive analyte recognition. Furthermore, ECL integration into μPADs-based and screen-printed electrodes have significantly increased the variety of options for analyte detection and proved perfect opportunities for biosensor fabrication (Wang 2013). Some of the latest examples of the paper-based BioMEMS used for electrochemiluminescence detection are as provided as an extension of this section.

Wang et al. (2013) developed a sensitive POC device for detection of carcinoma antigen 125 (CA125). The platform was composed of a 3D microfluidic origami device coupled with ECL immunosensor. Wax-printing was used for fabrication of this portable microfluidic origami device. Via screen printing counter electrodes and carbon working in a direct manner in addition to Ag/AgCl reference electrode created with their conductive pads over wax-patterned pure cellulose paper. These electrodes were triggered by folding the papers to create a 3D electrochemical cell (Fig. 8.1). The structure of this three-electrode system eliminates the potential influ-

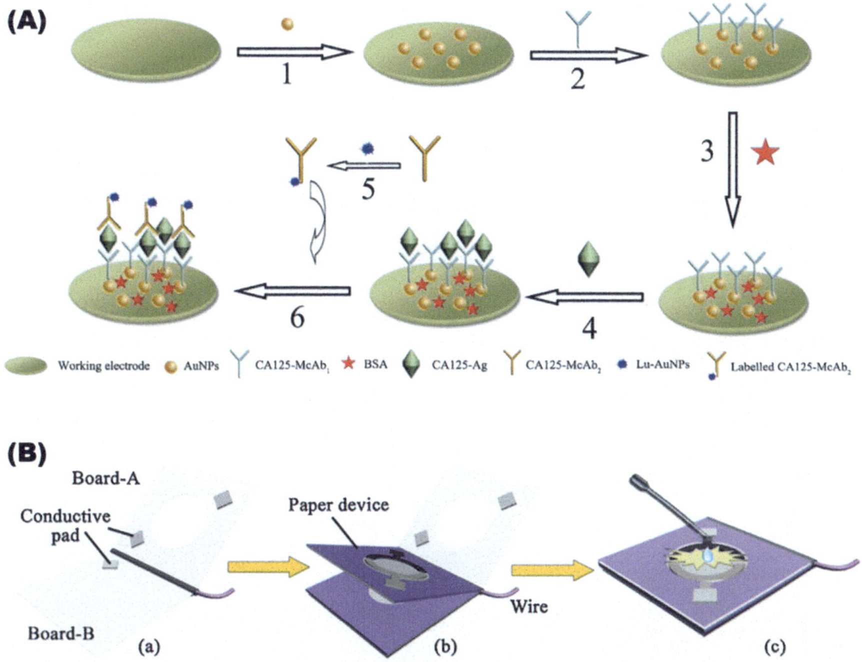

Fig. 8.1 **a** Schematic illustration of the fabrication of the ECL immunosensor and assay procedure; **b** Schematic illustration of the integration of 3D microfluidic origami immune-device with transparent device-holder (Wang 2013)

ence of reference, counter, and working electrodes on each other and minimizes the chance of contamination or damage to the working electrode during the operations of the immunoreactions. The authors enhanced the sandwich ECL immunosensor by employing luminol-functionalized gold nanoparticles (Lu-AuNPs) and gold nanoparticles (AuNPs), creating an immunosensor with low detection limit and significant sensitivity (0.0074 U mL^{-1} and 0.01–100 U mL^{-1}, respectively). To modify the working electrode while providing an excellent pathway of electron transfer, AuNPs was synthesized. The immobilized amount of capture antibody ($McAb_1$) on the working electrode was also enhanced. Furthermore, Lu-AuNPs was synthesized to mark the signal antibody ($McAb_2$). After the folding process, the developed ECL immunosensor was placed inside a device-holder prepared for the detection of CA125. For the ECL detection cyclic voltammetry was employed (Fig. 8.1). The proposed ECL immune-device was a simple, low-cost, disposal, and portable device with high level of sensitivity and considerably low limit of detection (LOD). The device has potentials for application if remote/rural areas (Wang 2013).

For the first time, a pen-on-paper electrochemiluminescence (PoP-ECL) apparatus was developed by Yang et al. (Table 8.1). It was entirely written and hand drawn. In order to produce an ECL immunosensor for detecting carbohydrate antigen 199

Table 8.1 Recent BioMEMS platforms for electrochemiluminescence detection: Type of the platform, main components, fabrication strategy, mechanism of operation. The analyte of interest, and the advantages and disadvantages of each platform are presented in this table

BioMEMS platform	Main components	Fabrication strategy	Mechanisms of operation	Detected analyte	Specifics	References
3D microfluidic origami ECL immunodevice	• Microfluidic paper-based platform • Screen-printing electrodes • Carbon ink • Cellulose paper	The origami apparatus was fabricated on pure cellulose paper by wax printing and patterning. On each wax-patterned paper, there were two circular paper working zones for screen-printing electrodes. In one circle, to screen-print the working electrode, carbon ink was utilized, and in another circle, carbon ink and Ag/AgCl ink were utilized accordingly, for screen-printing reference electrode and half-ring like counter electrode. The contact pads and conductive wires were screen-printed in the specified area and rectangular shape was achieved by cutting the paper sheet which were then folded to obtain the 3D microfluidic origami device	Lu-AuNPs labeled with $McAb_2$ were dropped onto the working electrode and allowed to dry. The device was folded and incorporated with a transparent device-holder to link with electrochemical system, which was composed of two circuit boards with conductive pads. To activate the ECL reaction, cyclic voltammetry measurements were used and a photomultiplier tube was implemented to record the light emissions	CA125	Signal amplification using Lu-AuNPs enhanced the detection results of sandwich-type ECL immunosensor with high sensitivity and low detection limit	Wang 2013

(continued)

Table 8.1 (continued)

BioMEMS platform	Main components	Fabrication strategy	Mechanisms of operation	Detected analyte	Specifics	References
PoP-ECL	• Cellulose paper • Commercial crayon • Rechargeable battery • 6B-type black pencil • Gold nanoparticles	A crayon was used to hand-draw hydrophobic regions on the paper, and then the wax-patterned paper sheet was baked to melt the wax and allow it to punctured through the paper. Furthermore, a 6B-type black pencil was used to hand-write conductive pads and PoP electrodes. Lastly, chitosan was used to cover the electrodes and to immobilize antibodies on the regions	To carry out the immunoreaction, sample solution with distinct concentrations of CA 199 in PBS was added PoP-ECL immune device followed by addition of labeled CA 199- $McAb_2$. For ECL detection, the PoP-ECL immune apparatus was linked with the power supply that was powered with the rechargeable battery	CA199	The device is low-cost, user-friendly, disposable, and portable	Yang 2014
Au-PWE ECL DNA sensor	• μPAD DNA sensor • GR/Au-PWE • $CaCO_3$/CMC hybrid microspheres • Luminescent AgNPs	The porous Au-PWE was altered by positively charged PPDDA-GR. Briefly, PDDA-GR solution was dumped onto the AuPWE, followed by washing. Using the DNA capture probe, the modified working electrode was incubated and a variety of target DNA concentrations were used for further incubation. Afterwards, reporter probe (S_3–$CaCO_3$/CMC@ AgNPs bioconjugates) was used to hybridize the electrode	Before the ECL measurement, the sample tab was folded down below the auxiliary tab and clamped into a home-made device-holder. The positive PDDA-GR was covered over the gold paper working electrode in combination with the microspheres which were covalently linked to ssDNA	DNA	The device was highly reliable and reproducible for DNA detection	Li et al. 2014

(continued)

Table 8.1 (continued)

BioMEMS platform	Main components	Fabrication strategy	Mechanisms of operation	Detected analyte	Specifics	References
μCAD	• White plain weave cotton cloth • Solid wax • CNB-7 • Biopotentiostat • Portable ECL measurement and analysis system	The fabrication of hydrophilic cloth chambers was performed through wax screen-printing technique was applied, and the electrodes were fabricated onto the cloth chamber through utilization of carbon ink. The cloth was placed on a heating board to form the hydrophobic barriers. Finally, the SPEs-containing cloth was isolated from the electrode-screen followed by drying	The μCAD device was placed inside a black box. The ECL chamber was aligned with a CCD camera and was connected to a PC for further analysis and control	Glucose	Low cost, disposable, rapid response rate, and portable total analysis system were the main advantageous features of this device	Guan et al. 2015)
Micro disk array-based biosensor	• GDAE • CQDs-AuNPs	The gold disk array working electrodes were immersed in a cysteamine solution to acquire the amine functionalized surface. The CQDs-AuNPs nano-hybrids solution was then drop-casted onto the amine-functionalized gold electrode surfaces	Cyclic voltammetry was employed to probe the modifications of the bare gold disk array electrode following CQDs/AuNPs-GOx layer creation	Glucose	The device showed a 13-fold rise in sensitivity of the detection compared to the counterpart planar biosensor	Buk and Pemble 2019)

(continued)

Table 8.1 (continued)

BioMEMS platform	Main components	Fabrication strategy	Mechanisms of operation	Detected analyte	Specifics	References
Cell-on-a-Chip (COC)	• Microdisc Electrode Array • Ag/AgCl reference electrode • Platinum mesh counter electrode	The device was fabricated by lithography technique and the electrochemical transducers were fabricated from electron beam vapor deposited platinum on an adhesion promoting Ti/W layer onto thick electronics grade borosilicate glass	A competitive ELISA was conducted using OTA and OTA-HRP solutions. The devices were dipped into the luminol-hydrogen peroxide solution and the amperometric response was recorded continually by the potentiostat and periodically by the luminometer	Ochratoxin A	The device presented a simultaneous amperometric and chemiluminescence recordings	Tria et al. 2016)
Electrochemical immunosensor	• Silanized silicon mold platform • PDMS coating, polyclonal antibodies for antigen measurement • MNPs • μWEs	The electrodes were modified by electro-addressing diazonium salt and electrodepositing magnetic nanoparticles coated with (Py/Py-COOH/MNPs). The Py/Py-COOH/MNPs coated core-shell magnetic nanoparticles were synthesized through utilization of a seeded-polymerization technique	TC immobilization on the μWEs surface was performed by functionalization with CMA. Activated electrode surfaces were incubated with TC. Via a second functionalization method, CMA was used for the activation of the μWEs followed by incubation with MNPs coated Py/Py-COOH/MNPs cross-linked with Ab-TC	Tetracycline	The immunosensor was capable of dramatically decreasing the time of analysis	Wang et al. (2018)

(continued)

Table 8.1 (continued)

BioMEMS platform	Main components	Fabrication strategy	Mechanisms of operation	Detected analyte	Specifics	References
Ni-GaIn strain sensor	• Liquid metal amalgams • Silicon rubber • Cooper wire	Through mixing Ni particles into gallium-based liquid metals (LMs), an enhanced adhesion was induced that allowed functional Ni-GaIn material to be printed onto a variety of soft substrates directly, to fabricate flexible electronics. The authors benefited from enhanced adhesion of Ni-GaIn amalgams, excellent softness, and high electro-conductivity, to form highly luminous and stretchable platforms	By applying changes in strain and force, the changes in the resistance were registered	Human body motions	This sensor system offers prospective applications in healthcare, prosthetic control, and soft robotics in the near future	Dickey (2017)

4-aminophenylacetic acid (CMA); Au-paper working electrode (Au-PWE); Calcium carbonate/carboxymethyl chitosan ($CaCO_3$/CMC); Carbohydrate antigen 199 (CA 199); Carcinoma antigen 125 (CA125); Conductive carbon ink (CNB-7); Electrochemiluminescence (ECL); Electrochemiluminescence (ECL); Gold microelectrodes (μWEs); Gold nanoparticles (AuNPs); Graphene-modified porous Au-paper working electrode (GR/Au-PWE); Luminol-functionalized gold nanoparticles (Lu-AuNPs); Magnetic nanoparticles (MNPs); Microcontact printing (μCP); Microfluidic cloth-based analytical devices (μCADs); Ni-doped liquid metal (Ni-GaIn); Ochratoxin A (OTA); Pen-on-paper electrochemiluminescence (PoP-ECL); Poly(pyrrole-copyrrole carboxylic acid) (Py/Py-COOH/MNPs); Polydiallyldimethylammonium chloride-functionalized graphene (PDDA-GR); Pulsed chronoamperometric technique (PCA); Screen-printed electrodes (SPEs), Self-assembled monolayers (SAMs); Silicon nitride (Si_3N_4); Silver nanoparticles (AgNPs); Single Stranded DNA (ssDNA); Tetracycline (TC); Titanium/tungsten (Ti/W)

(CA 199), a constant potential-triggered sandwich-type immunosensor was implemented into the PoP-ECL device. CA 199 is a popular tumor labeler utilized for pancreas and colon cancer diagnostic (Metzgar et al. 1982; Herlyn et al. 1982). To promote simplicity, the authors suggested to make use of a regular 6B-type black pencil to fabricate electrodes on the paper. The PoP-ECL apparatus composed of two PoP electrodes made up of commercially available pencil and caryon and a hydrophilic paper channel. For its fabrication, the hydrophobic regions were drawn by crayon on pure cellulose paper and the wax-covered paper was subsequently baked to form the hydrophilic paper channels. To perform precision writing of the carbon electrodes (named PoP electrodes) over the hydrophilic paper channel (Fig. 8.2), a 6B-type black pencil with low resistivity was employed. To create the immunosensor, chitosan was utilized to alter the paper working regions among two PoP electrodes to immobilize antibodies, covalently on the PoP-ECL device, while $Ru(bpy)_3{}^{2+}$ gold nanoparticles (Ru@AuNPs) were employed as the ECL luminophore. Additionally, to achieve constant-potential mode power supplier to the PoP electrodes to trigger the ECL for detection, a portable, low-cost rechargeable battery was employed. Several control experiments were carried out utilizing $Ru(bpy)_3{}^{2+}$ labeled CA 199- $McAb_2$, Ru@AuNPs labeled CA 199-$McAb_2$, and unmarked CA 199- $McAb_2$to as signal antibody to examine the ECL performance of Ru@AuNPs labeled CA 199-$McAb_2$

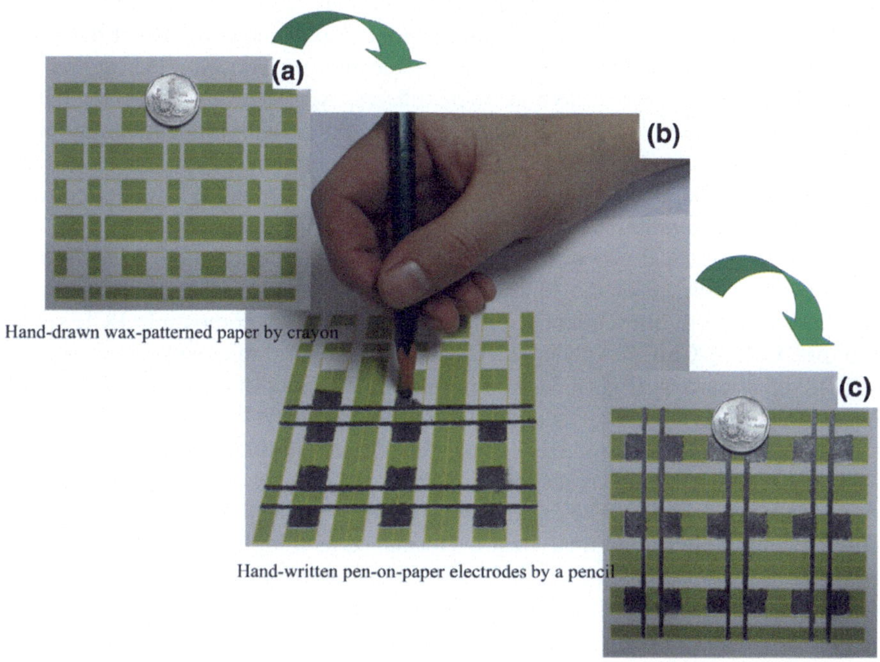

Fig. 8.2 Schematic rillustration of the fabrication of PoP-ECL device (Yang 2014)

and the emissions of ECL were compared. The results proved that PoP-ECL immunodevice showed an acceptable linear response range from 0.01 to 200 U mL^{-1} with a LOD of 0.0055 U mL^{-1}. Overall, the benefits of the suggested immunodevice include: (i) development of wax-covered paper by means of employing an inexpensive crayon; (ii) application of commercially available 6B-type black pencil for creating carbon electrodes on μPADs; (iii) integration of a rechargeable battery onto PoP-ECL device which significantly simplified the device and decreased the cost (Yang 2014).

Using a compatibly designed origami electrochemical device, a unique porous Au-paper working electrode (Au-PWE) that coupled the high conductivity of AgNPs and porosity of paper, was fabricated. By an interrelated AuNPs layer on the fibers' sample zone, the conductivity of the paper was significantly enhanced. Li et al. (2014) created a sensitive and simple ECL-DNA sensor founded on calcium carbonate/carboxymethyl chitosan ($CaCO_3$/CMC) hybrid microspheres @ luminescent AgNPs composites and graphene-modified porous Au-paper working electrode (GR/Au-PWE) (Table 8.1). Carboxymethyl chitosan (CMC) was utilized to control the particle size and to form the hybrid particles ($CaCO_3$/CMC) with great biodegradability and biocompatibility properties, fine loading capability and substantial specific surface area. These particles were developed through precipitating the calcium carbonate in an aqueous solution including CMC implemented as carriers for immobilization of AgNPs. Thermal reduction of silver ions in a glycine matrix was performed to synthesize the AgNPs, making use of the solid-state matrix to manage the migration and nucleation of reduced silver atoms. On the surface of $CaCO_3$/CMC hybrid microspheres, AgNPs were covalently bound to the complementary ssDNA sequence. Positive poly (diallyldimethylammonium chloride)-functionalized graphene (PDDA-GR) was compunded to Au-coated cellulose fibers in the paper sample zone, in which PDDA-GR/Au-PWE was effectively developed for the immobilization of capture probe, in order to further enhance the Au-PWE's electrochemical properties. Finally, the $CaCO_3$/CMC@AgNPs labels were brought to the surface of the PDDA-GR/Au-PWE through subsequent sandwich DNA hybridization. Subsequently, to acquire a unique biocompatible ECL signal amplifier, $CaCO_3$/CMC@AgNPs, the $CaCO_3$/CMC hybrid microspheres were used as ECL signal carriers (Fig. 8.3). Making use of dual amplification impacts of the $CaCO_3$/CMC@AgNPs composites and GR modified Au-PWE, the target DNA could be detected by the paper-based DNA sensor, quantitatively, in the range of 4.0×10^{-17} to 5.0×10^{-11} M, showing great specificity, with a LOD as low as 8.5×10^{-18} M, and with excellent selectivity. The experimental results indicated that the μPADs DNA sensor exhibited appealing analytical performance as a high-throughput, portable, rapid, low-cost and simple ECL μPADs. It is a favourable platform for precise gene diagnostics on site and at home that could be readily applicable for POC testing, environmental monitoring, and public health in remote regions (Li et al. 2014).

The initial attempt at merging ECL detection and micro cloth-based analytical devices (μCADs) to measure glucose was presented by Guan et al. (2015). The integration of ECL assay with μCADs can provide tremendous benefits including portability, cost-effectiveness and simplicity of the devices in addition to high selectivity

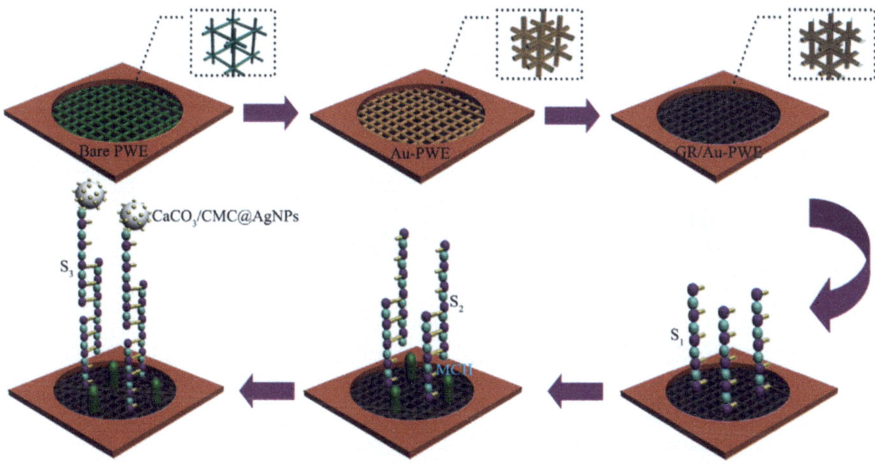

Fig. 8.3 Schematic illustration of the fabrication procedures for the μPAD DNA sensor (Li et al. 2014)

and sensitivity of the detection approach. The ECL device development was made simple, low-cost, and rapid, through adopting two screen-printing processes, while remaining suitable and effective for fabricating vast range of devices. A newly demonstrated wax-screen-printing technology (Guan et al. 2015) was used for the fabrication of the cloth-based microfluidic ECL chamber. A facile carbon ink-screen-printing was utilized to pattern the chamber with carbon screen-printed electrodes (SPEs), to fabricate three carbon electrodes directly on the patterned cloth (Fig. 8.4) (Guan et al. 2015). Using this method, the wax-screen-printed hydrophilic cloth patterns were protected from the exposure towards organic solvents or to photoresists that could interfere with the reaction on cloth or contaminate the platform. Moreover, the ECL μCADs was able to conduct quantitative detection of target analytes, tris(2,2′bipyridyl) ruthenium(II)/tri-n-propylamine ($Ru(bpy)_3^{2+}$/TPA) and 3-aminophthalhydrazide/H_2O_2 (luminol/ H_2O_2), as the ECL intensity was recorded through a portable CCD-based imaging system. The ECL assays utilizing luminol/ H_2O_2 and $Ru(bpy)_3^{2+}$/TPA systems could be carried out, where the fundamentals are shown in Fig. 8.5a, and the illustration of the analysis system for cloth-based microfluidic ECL and portable imaging, is shown in Fig. 8.5b. A rapid ECL response rate was indicated when the μCADs used $Ru(bpy)_3^{2+}$/TPA and luminol/ H_2O_2 reaction systems, and the assay procedure was completed in 30 s. Moreover, the detection limits of 0.027 mM for H_2O_2 and 1.265 μM for TPA detection were shown by the μCADs. Additionally, the applicability of this device was shown for determining glucose in artificial urine (AU) samples and PBS solution with the LOD of 0.038 and 0.032 mM, respectively. The results of this study show that μCADs is a great candidate from various sensing and biosensing applications (Guan et al. 2015).

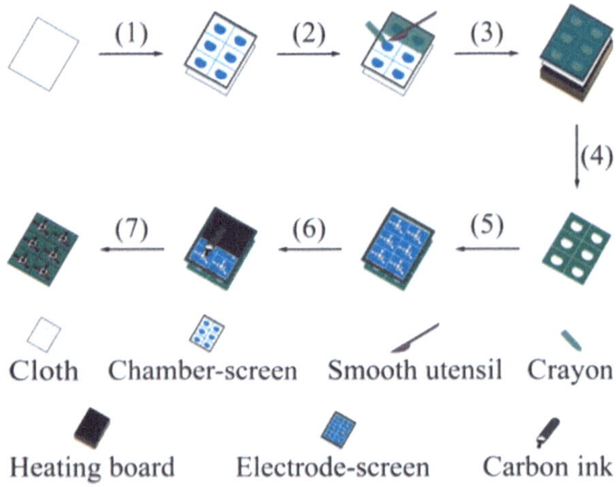

Fig. 8.4 Schematic representation of screen-printing process for fabricating μCADs. (1) A chamber-screen is positioned on the cloth. (2) The chamber-screen is rubbed with a smooth utensil and a crayon, in turn. (3) The cloth is positioned on the heating board along with the chamber-screen. (4) The cloth and chamber-screen are detached from the heating board and isolated from each other to create wax-patterned cloth. (5) An electrode-screen is positioned on the patterned cloth. (6) Carbon ink is screen-printed onto the cloth. (7) The resultant cloth is isolated from the electrode-screen, and dried at room temperature (Guan et al. 2015)

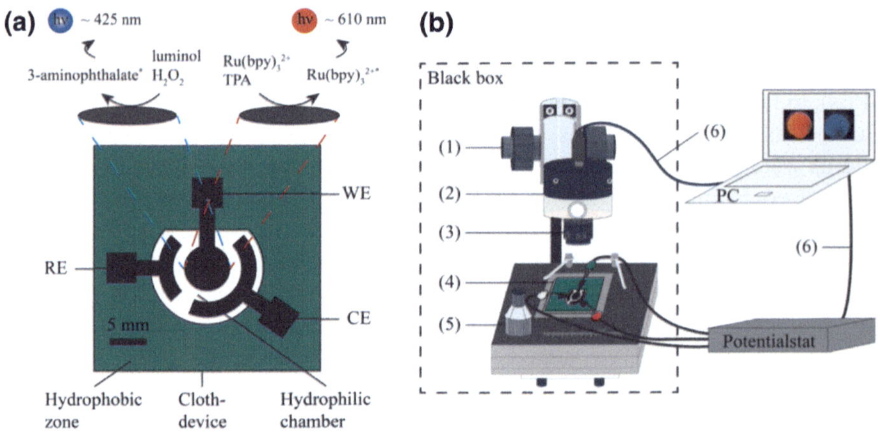

Fig. 8.5 Schematic diagram of ECL assays on μCADs. **a** The assay principles and designed sizes of the device. In the top view, the $Ru(bpy)_3{}^{2+}$/TPA- and luminol/H_2O_2-based ECL assay principles are shown, while the bottom view shows the design and sizes of the device (WE—working electrode, CE—counter electrode, RE—reference electrode). **b** Schematic representation of the portable imaging and analysis system for cloth-based microfluidic ECL: (1) focus-adjustable bracket; (2) CCD; (3) macro-lens; (4) cloth device-containing plastic support; (5) movable mechanical stage; and (6) USB interface line (Guan et al. 2015)

8.4 Alternative BioMEMS for Electrochemiluminescence Detection

Arguably, the most crucial aspect of biosensor development is the immobilization of the biorecognition or biomolecular elements onto an appropriate matrix. Alternation of the surface with biocompatible polymers, carbon-based nanomaterials, or metal nanoparticles are amongst the most frequent ways to form selectively favorable surfaces for biomolecules (Holzinger et al. 2014; Buk et al. May 2017; Derkus et al. 2014, 2015). Micro disk array electrodes represent a commonly studied microelectrode geometry for creating a steady-state current level (Buk and Pemble 2019). In a novel study, electronics-standard lithography, deposition, and etching methods were implemented to microfabricate gold micro disk array electrodes (GDAE) on Si substrate. The electrodes were fabricated by utilizing standard Si/SiO_2/metal microfabrication technology. Through this technology, to perform the patterning procedure for the disk arrays the etching of a passivation layer was required that was deposited on top of the metal layer. Each individual microelectrode composed of 85 gold disk electrodes having 200 mm inter-electrode distance and 20 mm diameter and were positioned hexagonally. A hybrid nano-material containing two distinctive types of nanoparticles, carbon quantum dots (CQDs) and AuNPs were employed to modify the gold electrode surfaces (Table 8.1). These particles offer favorable chemical and physical characteristics which were used for fabrication of the miniaturized biosensor for glucose detection. The electrodes were characterized electrochemically to examine the microfabrication route's efficacy. Moreover, several immobilization methods were performed in order to prepare CQDs/AuNPs-GOx micro disk array electrodes (Fig. 8.6). The principal components of the assay consist of GOx (immobilized on the nanoparticles), phosphate buffer saline tablets, glucose, potassium chloride, sodium chloride, sulphuric acid, potassium ferrocyanide, cysteamin, acetaminophen, and uric acid. The authors successfully developed a highly reliable and reproducible electrochemical biosensor that merges the utilization of the miniaturized electrode technologies and CQDs/AuNPs nanohybrid materials. This platform is suited for future development along the path of achieving an entirely on-chip system, with potential further miniaturization, making it especially appealing for a variety of applications (Buk and Pemble 2019).

NFO4 was used as the immobilized synthetic peptide to develop an amperometrics biosensor with high binding affinity. This study presented an application of dual detection techniques applied into an electrochemical cell-on-a-chip (ECC) microdisc electrode array, which showed great promises for further developments of biosensors. The electrochemical ECC and bio-transducers were fabricated by lithography technique. A microdisc electrode array working electrode format was used for this technique accompanied with microporous graphitized carbon (MGC) electrodeposited inside a poly (aniline-co-meta-aminoaniline) electroconductive polymer layer (Table 8.1). As a redox mediator, iron-nickel hexacyanoferrate (Fe|NiHCF) was deposited onto the MGC, amperometrically. Furthermore, the use of a Ni-FeHCF layer enhanced the peroxide signal that would have been otherwise compromised

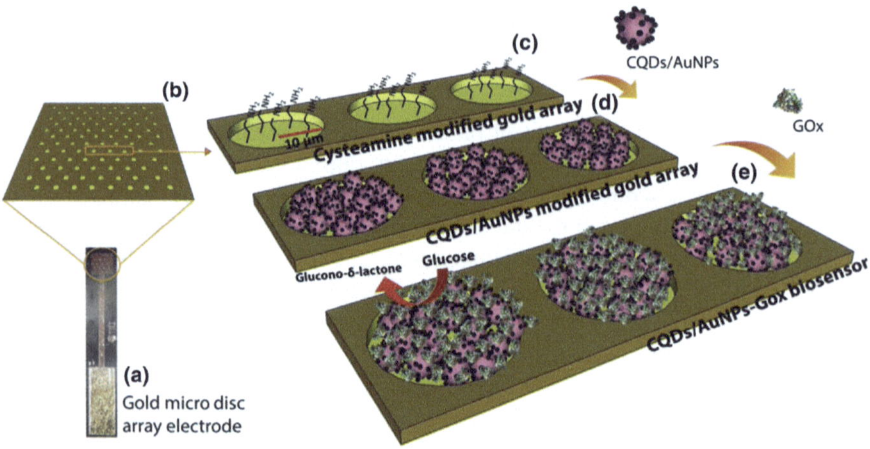

Fig. 8.6 Schematic representation of the biosensor development process; **a** the bare single disk array electrode, **b** array's magnified surface, **c** amine functionalized gold surface following cysteamine alteration, **d** CQDs/AuNPs attached surface, **e** GOx enzyme immobilized entire CQDs/AuNPs-GOx biosensor. Consider that the size of the biomolecule, nanomaterials and electrodes indicated are not drawn to scale (Buk and Pemble 2019)

when an electropolymerized polyaniline layer was first deposited to support hydrogel attachment. The apparatus was dip-coated with monomer cocktail that produced poly (2-hydroxyethyl methacrylate-co-2-aminoethyl methacrylate) or poly(HEMA-co-AEMA). This polymer foam was formed by UV crosslinking creating a 3-D support for the chelation of Zn^{2+} ions ($ZnCl_2$) and the subsequent immobilization of N-terminus his-tagged peptide, NFO4. The newly developed biosensor was used for molecular recognition of ochratoxin A (OTA), a natural carcinogenic mycotoxin that simulates the mycotoxin-specific antibody (Fig. 8.7). For the assay, horseradish peroxidase (HRP) conjugated OTA was combined with the OTA solution which were incubated together on the biospecific MDEA ECC 5037-Pt|MGC|HCF|Hydrogel-NFO4 bio-transducer, competitively. After the addition of H_2O_2/luminol substrate, the amperometric response to peroxide was measured. A concurrent analysis of light emission signals enabled the opportunity to directly compare the performance of chluminescence and amperometric. These performances were found comparable in their dynamic range and detection limits (Tria et al. 2016).

El Alami El Hassani et al. (2019) developed a novel BioMEMS immunosensor based on an integrated transducer including eight gold microelectrodes (μWEs) as well as counter electrodes, integrated silver and platinum reference (Table 8.1) (El Alami El Hassani 2019). The electrodes were modified by electro-addressing diazonium salt and poly (pyrrole-co-carboxylic acid) (Py/Py-COOH/MNPs) coated electrodepositing magnetic nanoparticles. Py/Py-COOH/MNPs was used to coat the core-shell magnetic nanoparticles which were then synthesized through utilization of seeded-polymerization technique. The immobilization and functionalization processes of μWEs were two main elements that contributed to the novelty of this

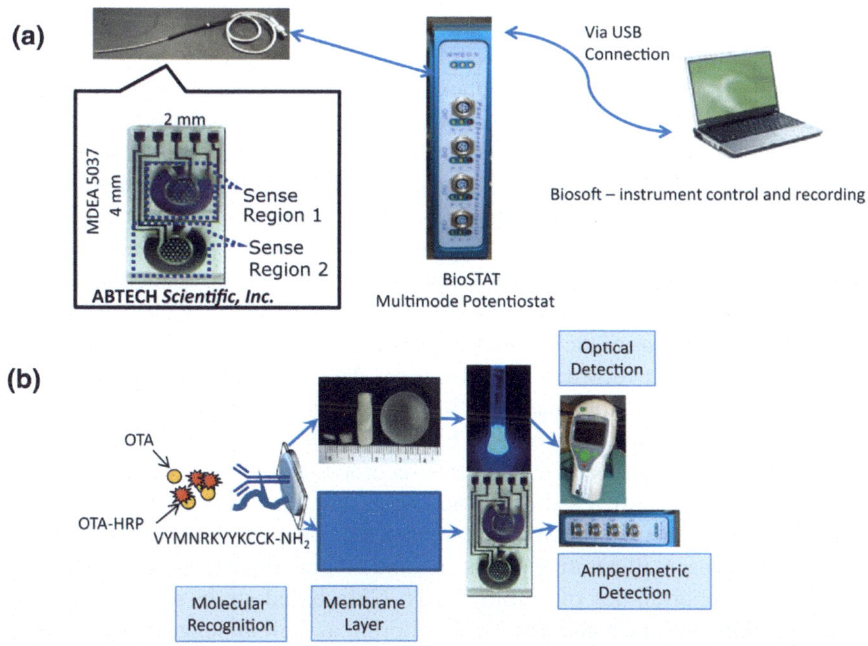

Fig. 8.7 **a** Micrograph of the two-channel electrochemical transducer, the MDEA 5037-Pt, and its interfacing to the four-channel multimode potentiostat, the esa Biostat, wherein two channels are reconfigured to a single connector port.**b** Schematic layout for the comparative evaluation of the two biosensor systems using the same molecular recognition elements. Above: Optical detection mode (luminescence) using a luminometer. Below: Amperometric detection mode using a multimode potentiostat (Tria et al. 2016)

work as follows: (i) A diazotized CMA solution in an aqueous solution of HCl and $NaNO_2$ was used to functionalize the μWes; (ii) nine-repetitive cyclic voltammograms were employed and the resultant carboxylic groups created on the electrode surfaces were triggered through carbodiimide chemistry; (iii) the triggered electrode surfaces were incubated in tetracycline (TC) solution and were further treated with ethanolamine in phosphate buffer saline (PBS) buffer to prevent nonspecific bonding. Subsequently, purified TC polyclonal antibody (Ab-TC) was combined with the triggered nanoparticles. The target analyte was measured through competitive detection of Ab-TC and TC, by implementing a combination of decreasing concentrations of TC and a fixed concentration of Ab-TC. The results showed that the immunosensor was highly sensitive with LOD of 1.2 pg mL^{-1} while providing high reproducibility and rapid response. This platform could dramatically decrease the time of analysis providing a new pathway for advanced immunoassays development in industrial food control (El Alami El Hassani 2019).

Wang et al. (2018), developed a highly stretchable and conformable strain sensor fabricated by Ni-doped liquid metal (Ni-GaIn) designed to record and reconstruct human motion at elbow, knee, heel and even fingers for multifunctional human-activity monitoring (Table 8.1). Through mixing Ni particles into gallium-based liquid metals (LMs), an enhanced adhesion was induced that allowed fabrication of flexible electronics as the functional Ni-GaIn material was allowed to be printed onto a variety of soft substrates, directly (Dickey 2017). The authors benefited from enhanced adhesion of Ni-GaIn amalgams, appealing softness, and high electro-conductivity to establish highly luminous and stretchable platforms made from Ecoflex 00-30 encapsulation layer (Ecoflex), Ecoflex 00-30—ZnS substrate layer (Ecoflex—ZnS) and Ni-GaIn wire. Ultimately, a sensing glove founded on Ni-GaIn sensor was developed to oversee the local movement, where they evaluate its electrical stability by means of the correlation between deformation and resistance value (Wang et al. 2018). It recorded the movement data and reconstructed the motion of the fingers through implementing a human–machine interface in the computer, in which, as the Ni-GaIn sensor was bent, stretched, and compressed, the electrical signals were measured. The authors recorded fatigue tests where the device illustrated perfect electrical stability when being used as a sensor. The device was capable of concurrent tracking of the motions of the fingers through utilization of five distinct strain sensors illustrating precise readout of complicated motions. The study indicated the opportunities of employing Ni-GaIn based strain sensor system for human–machine interface and activity monitoring as well as suggesting substantial prospective applications for human motion quantification (Wang et al. 2018).

8.5 Summary

By integrating the sensitivity of chemiluminescence with advantageous features of electrochemistry, electrochemiluminescence offers a sensitive, nonhazardous, and inexpensive method for biorecognition of a great number of target analytes. In combination with BioMEMS, electrochemiluminescence can generate powerful biosensing platforms that are benefited from portability, compactness, cost-effectiveness, and ease of signal read out. Such devices can serve as great candidates in normal urban settings for point of care or in remote and/or rural areas for extreme point of care. This chapter presented a summary of the latest advancements of this hybrid systems that involve both BioMEMS and electrochemiluminescence detection strategy.

References

Bard AJ (1988) Electrogenerated chemiluminescence-final report

Blair EO, Corrigan DK (2019) A review of microfabricated electrochemical biosensors for DNA detection. Biosens Bioelectr 134:57–67. doi: https://doi.org/10.1016/j.bios.2019.03.055

Buk V, Pemble ME (2019) A highly sensitive glucose biosensor based on a micro disk array electrode design modified with carbon quantum dots and gold nanoparticles. Electrochim Acta 298:97–105. https://doi.org/10.1016/j.electacta.2018.12.068

Buk V, Emregul E, Emregul KC (2017) Alginate copper oxide nano-biocomposite as a novel material for amperometric glucose biosensing. Mater Sci Eng C 74:307–314. https://doi.org/10.1016/J.MSEC.2016.12.003

Davies R, Bartholomeusz DA, Andrade J (2003) Personal sensors for the diagnosis and management of metabolic disorders. IEEE Eng Med Biol Mag 22(1):32–42. https://doi.org/10.1109/MEMB.2003.1191447

Derkus B, Emregul E, Emregul KC, Yucesan C (2014) Alginate and alginate-titanium dioxide nanocomposite as electrode materials for anti-myelin basic protein immunosensing. Sens Actuat B Chem 192:294–302. https://doi.org/10.1016/J.SNB.2013.10.128

Derkus B, Emregul KC, Emregul E (2015) Evaluation of protein immobilization capacity on various carbon nanotube embedded hydrogel biomaterials. Mater Sci Eng C 56:132–140. https://doi.org/10.1016/J.MSEC.2015.06.022

Dickey MD (2017) Stretchable and soft electronics using liquid metals. Adv Mater 29(27):1606425. https://doi.org/10.1002/adma.201606425

Farokhi-Fard A, Golichenari B, Ghanbarlou MM, Zanganeh S, Vaziri F (2019) Electroanalysis of isoniazid and rifampicin: role of nanomaterial electrode modifiers. Biosens Bioelectron 111731. doi: https://doi.org/10.1016/j.bios.2019.111731

Guan W, Liu M, Zhang C (2015) Electrochemiluminescence detection in microfluidic cloth-based analytical devices. Biosens Bioelectron 75:247–253. https://doi.org/10.1016/j.bios.2015.08.023

Guan W, Zhang C, Liu F, Liu M (2015) Chemiluminescence detection for microfluidic cloth-based analytical devices (μCADs). Biosens Bioelectron 72:114–120. https://doi.org/10.1016/J.BIOS.2015.04.064

El Alami El Hassani N et al (2019) Development and application of a novel electrochemical immunosensor for tetracycline screening in honey using a fully integrated electrochemical Bio-MEMS. Biosens Bioelectron 130: 330–337. doi: https://doi.org/10.1016/j.bios.2018.09.052

Herlyn M, Sears HF, Steplewski Z, Koprowski H (1982) Monoclonal antibody detection of a circulating tumor-associated antigen. I. Presence of antigen in sera of patients with colorectal, gastric, and pancreatic carcinoma. J Clin Immunol 2(2):135–140. https://doi.org/10.1007/BF00916897

Holzinger M, Le Goff A, Cosnier S (2014) Nanomaterials for biosensing applications: a review. Front Chem 2. doi: https://doi.org/10.3389/fchem.2014.00063

Hu L, Reviews GX-CS et al (2010) Applications and trends in electrochemiluminescence. pubs.rsc.org

Ju H, Lai G, Yan F (2017) Electrochemiluminescent immunosensing. In: Immunosensing for detection of protein biomarkers. Elsevier, Amsterdam, pp 171–206

Li M, Wang Y, Zhang Y, Yu J, Ge S, Yan M (2014) Graphene functionalized porous Au-paper based electrochemiluminescence device for detection of DNA using luminescent silver nanoparticles coated calcium carbonate/carboxymethyl chitosan hybrid microspheres as labels. Biosens Bioelectron 59:307–313. https://doi.org/10.1016/j.bios.2014.03.072

Li C, Wang Z, Wang L, Zhang C (2019) Biosensors for epigenetic biomarkers detection: a review. Biosens Bioelectron 144:111695. https://doi.org/10.1016/j.bios.2019.111695

Metzgar RS, Gaillard MT, Levine SJ, Tuck FL, Bossen EH, Borowitz MJ (1982) Antigens of human pancreatic adenocarcinoma cells defined by murine monoclonal antibodies. Cancer Res 42(2):601–608

Pfeiffer SA, Borisov SM, Nagl S (2017) In-line monitoring of pH and oxygen during enzymatic reactions in off-the-shelf all-glass microreactors using integrated luminescent microsensors. Microchim Acta 184(2):621–626. https://doi.org/10.1007/s00604-016-2021-2

Ramon C, Temiz Y, Delamarche E (2017) Chemiluminescence generation and detection in a capillary-driven microfluidic chip . Microfluid BioMEMS Med Microsyst XV 10061:100610O. https://doi.org/10.1117/12.2250765

Richter MM (2008) Electrochemiluminescence. In: Optical biosensors. Elsevier, pp 317–384

Roda A et al (2017) Smartphone-based biosensors for bioanalytics: a critical review. Compr Anal Chem 77:237–286. https://doi.org/10.1016/bs.coac.2017.05.007

Socorro-Leránoz AB, Santano D, Del Villar I, Matias IR (2019) Trends in the design of wavelength-based optical fibre biosensors (2008–2018). Biosens Bioelectron X 1:100015. https://doi.org/10.1016/J.BIOSX.2019.100015

Tria SA, Lopez-Ferber D, Gonzalez C, Bazin I, Guiseppi-Elie A (2016) Microfabricated biosensor for the simultaneous amperometric and luminescence detection and monitoring of Ochratoxin A. Biosens Bioelectron 79:835–842. https://doi.org/10.1016/j.bios.2016.01.018

Wang S et al (2013) 3D microfluidic origami electrochemiluminescence immunodevice for sensitive point-of-care testing of carcinoma antigen 125. Sens Actuat B 176:1–8. https://doi.org/10.1016/j.snb.2012.08.035

Wang X, Guo R, Yuan B, Yao Y, Wang F, Liu J (2018) Ni-doped liquid metal printed highly stretchable and conformable strain sensor for multifunctional human-motion monitoring

Yang H et al (2014) Hand-drawnampamp;written pen-on-paper electrochemiluminescence immunodevice powered by rechargeable battery for low-cost point-of-care testing. Biosens Bioelectron 61:21–27. https://doi.org/10.1016/j.bios.2014.04.051

Zhang X et al (2019) Recent advances in the construction of functionalized covalent organic frameworks and their applications to sensing. Biosens Bioelectron 145:111699. https://doi.org/10.1016/j.bios.2019.111699